David R. Dobrzykowski 9/85

$8.00

AF545071

Stratigraphic Modeling and Interpretation: Geophysical Principles and Techniques

Education Course Note Series # 13

Norman S. Neidell
Zenith Exploration Company
Houston, Texas

This book is an author-prepared publication
of the AAPG Education Program.
Extra copies of this, and other titles in the
Education Course Note Series, are available from:

The AAPG Bookstore
P.O. Box 979
Tulsa, Oklahoma 74101, USA
918-584-2555

First Printing September 1979
Fourth Printing March 1984

ISBN: 0-89181-162-1

Table of Contents

Table of Contents
(Cont'd.)

I. Relating Seismic Data to Stratigraphy

A. General Statements

Stratigraphic interpretation of seismic data requires that the seismic information be expressed in geological terms. The strictly geological view of the earth is developed from surface observations, guiding principles of geological evolution and subsurface information from bore holes. Amongst the subsurface information are well log measurements from a variety of physical sensors. We can correlate the information content of the seismic data with the geologic view in terms of geometric description and a subset of the log measurements, namely velocity and density. Such correlation used in context with geological and geophysical principles is at the heart of stratigraphic interpretation.

Typically available well log measurements include the Spontaneous Potential, Resistivities, Radioactivity parameters, Sonic Travel Times (Velocity) and Densities. While the vertical detailing of such measurements is excellent, they are quite limited in the extent to which they define subsurface parameters laterally. Hence, a strong interpretive element enters the picture in which appeal is made to fundamental geological concepts and principles. Still, such measurements can provide insights about the porosity and fluid contents of the rocks and the lithology in general which enhance stratigraphic conclusions based solely on a geologic approach.

Seismic measurements on the other hand do define in some sense the subsurface geometry and give estimates of the acoustic impedance which is related to the rock velocities and densities. The vertical detail is rather limited owing to the size of the seismic wavelet and it's observation principally as sequences of overlapped events, but the lateral definition is quite good although averaged over regions know as Fresnel zones. Also, specialized analyses and interpretive methods can make use of indirect information to provide insights into the nature of porosity, fluid content and lithology in general. Here again fundamental principles and strong interpretive elements are involved.

In the face of these backgrounds we may then ask for a specific vehicle by which the geologic view and the seismic information may be correlated. The most elementary form of such a vehicle has been available to us for some time now - the Synthetic Seismogram (see Peterson, Fillipone and Coker (1955) and Wuenchel (1960)).

The basis of the synthetic seismogram rests on a viewpoint of a laterally restricted segment of the earth taken as a horizontally layered medium. A plane wave is assumed to propagate in this earth and the seismogram is taken to be made up of the partial reflections of the plane wave from each of the planar boundaries (see Figure). Note that the change in lithologies across the boundary is represented by differing velocities and densities.

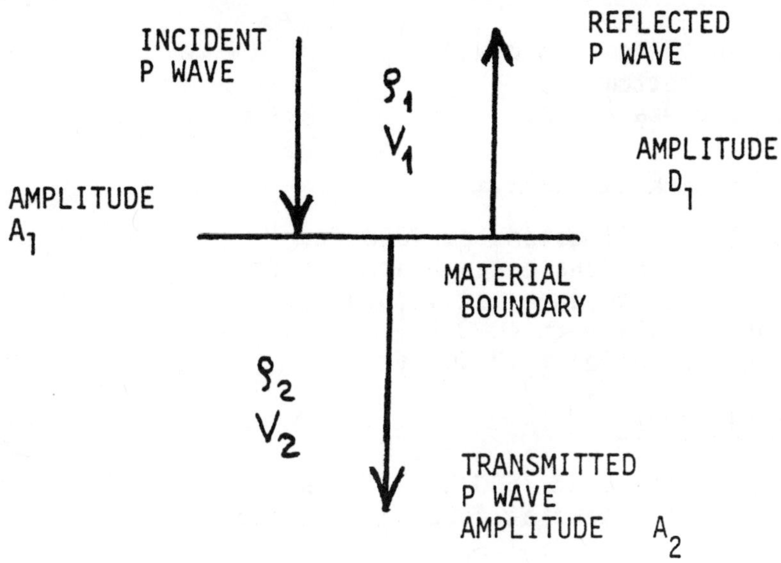

REFLECTION AND TRANSMISSION AT A LITHOLOGIC BOUNDARY

The product of the velocity and density of a material is known as the acoustic impedance, and the reflection coefficient and transmission coefficients which characterize the amplitudes of the reflection and transmitted wave are simply written in terms of such products. For the case illustrated, the reflection coefficient r and transmission coefficient t may be written as follows:

$$r = \frac{D_1}{A_1} = \frac{\rho_2 V_2 - \rho_1 V_1}{\rho_2 V_2 + \rho_1 V_1} \;; \qquad t = \frac{A_2}{A_1} = \frac{2\rho_1 V_1}{\rho_2 V_2 + \rho_1 V_1}$$

In particular we should recognize that r is most sensitive to the change in the acoustic impedance across the boundary rather than its specific magnitude. A short table of reflection coefficient values follows which can be most useful as a guide for understanding the relative amplitudes of different reflection events.

TYPICAL REFLECTION COEFFICIENTS

Near-surface reflectors:	Reflection coefficient
Soft ocean bottom (sand/mud)	0.33
Hard ocean bottom	0.67
Base of the weathering	0.63
Good or strong reflectors:	
Sand/shale vs. limestone at 4000 ft.	0.21
High/low of sand/shale range at 4000 ft.	0.14
Sand/shale vs. basement at 12,000 ft.	0.29
Gas sand vs. shale at 4000 ft.	0.23
Gas sand vs. shale at 12,000 ft.	0.125
Fair reflectors:	
Velocity contrast of 7500/8500 ft./sec.	0.06
Multiple of gas sand at 4000 ft. and ocean bottom	0.04-0.02

Reflection coefficients normally decrease with depth because the percentage of variation of the acoustic impedance (velocity x density) decreases with depth. Where nothing is known about the stratigraphy of an area, a limestone formation may easily be interpreted as a Gulf Coast type gas sand if only reflection magnitude is considered.

An example after Biggs, Blakely and Rudman (1960) showing the essentials of a geological correlation with seismic data in Lawrence County, Indiana is next illustrated. The lithologic units are identified and their seismic velocities as determined from well logs are shown. Similarly the lithologic units corresponding to the seismic expression are indicated. A synthetic seismogram developed from the velocity information is presented for correlation with the actual seismic data. The agreement in character is quite good, but on close inspection, the agreement in detail is less satisfactory.

By contrast, another illustration which follows over an oil field in a North Sea Jurassic sand shows a far more definitive correlation. In this instance, both velocity and density logs were used to form reflection coefficients and the waveform taken in developing the synthetic seismic data matches the propagating waveform of the seismic data. For this comparison the actual data has been separated at the trace closest to the well location. The agreement in detail here is remarkably good and we shall return to discuss other aspects of this correlation later. Still, two important lessons emerge. First, we may not neglect density as a physical parameter without degrading the effectiveness of our correlation and second, the agreement in waveform between the synthetic data and the real data is an essential ingredient of the correlation.

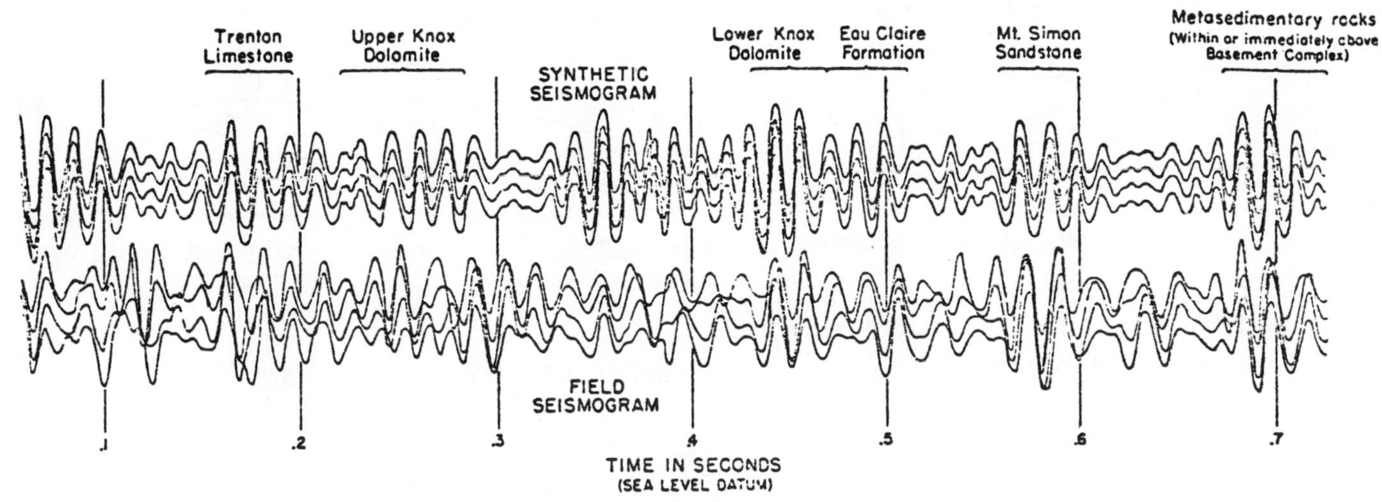

Synthetic and field seismograms from Lawrence County showing principal reflecting zones.

TEST WELL IN LAWRENCE COUNTY

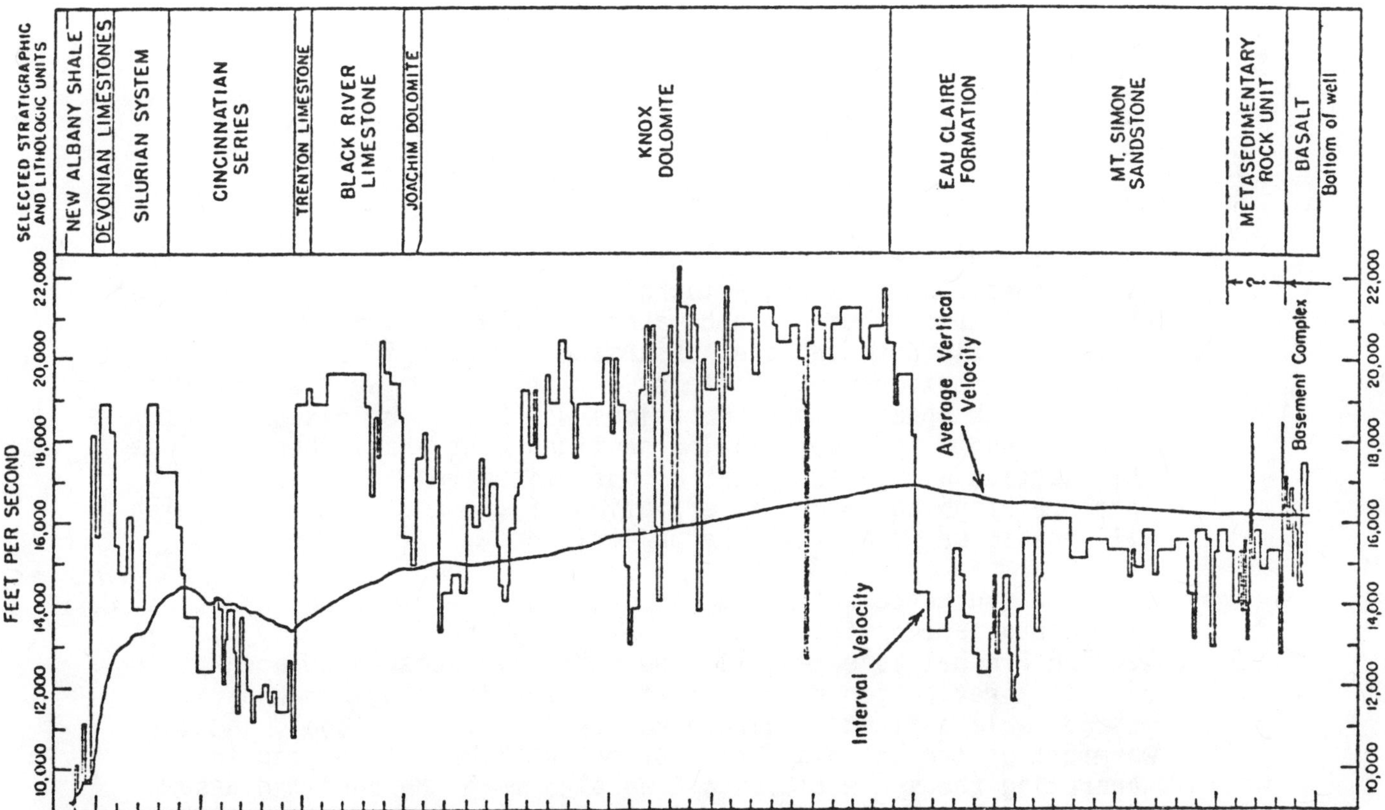

Interval and average vertical velocities as a function of depth.

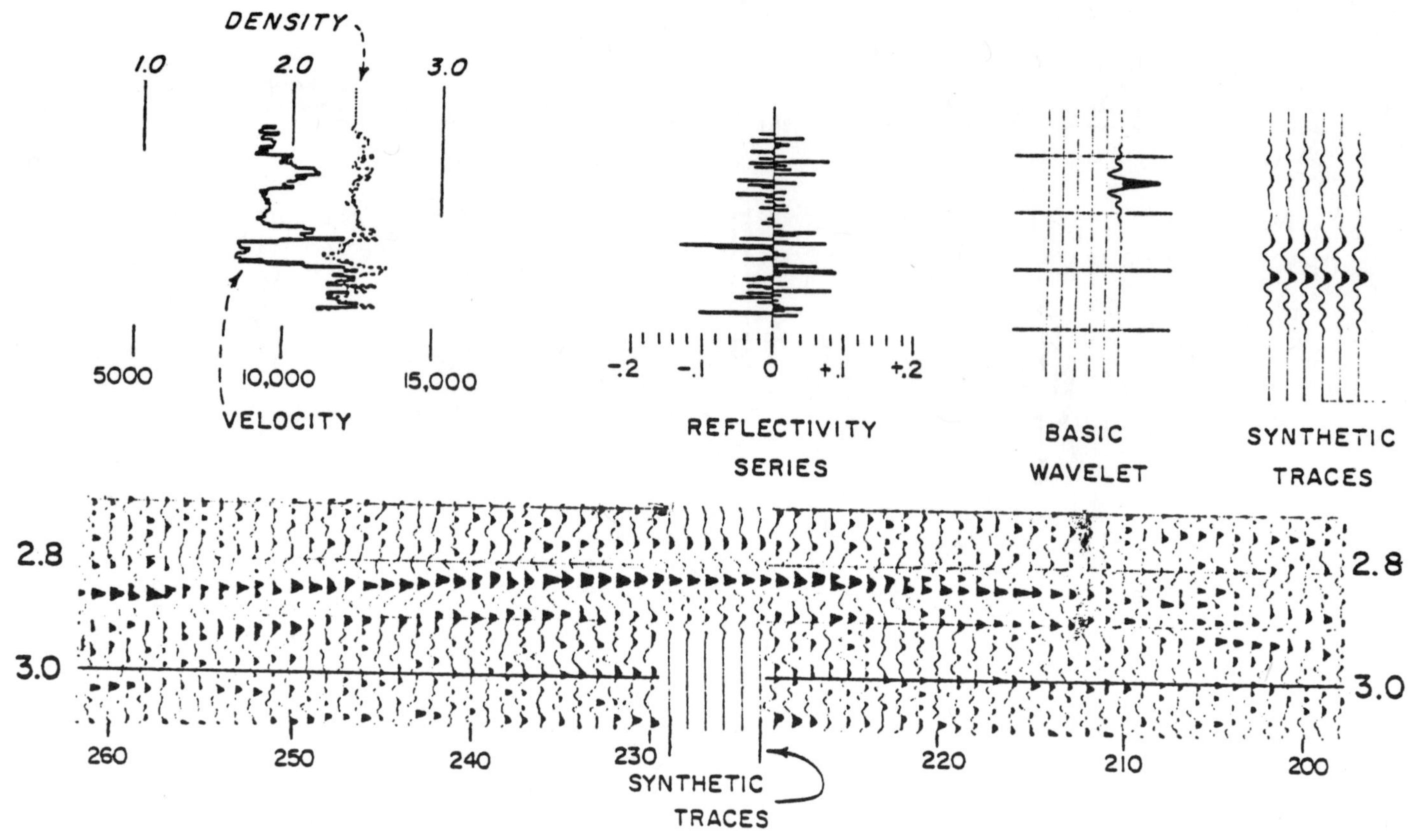

CORRELATION OF GEOLOGY WITH SEISMIC DATA

VIA THE SYNTHETIC SEISMOGRAM

Matching synthetic seismograms and processed seismic sections is of course, an elementary technique of stratigraphic interpretation which while easily understood makes a number of strict requirements on the seismic data and the subsurface and also ignores a number of important questions and problems. At this point we shall mention several such specific considerations which must be discussed at greater length in the following sections. We shall then conclude this section with a statement of our objectives in stratigraphic interpretation and a characterization of the analytic tools which will enable us to achieve the objectives.

The synthetic seismogram as described requires an essentially flat subsurface having lateral continuity and homogeneity over at least a Fresnel zone which is the effective subsurface area giving rise to a reflection event. Further, the lithologic boundaries must be well defined in nature rather than transitional, and the waveform of the seismic data must match the waveform used in generating the synthetic data. We also make the unstated assumption that each event on the seismic data corresponds to a lithological boundary and thus rule out the possibility of noise and other events of no direct geologic significance.

B. Departures of the Real World from the Ideal

Random noise is well known to the seismic interpreter as is the lack of significance of such events. These events are recognized principally by their lack of coherence or continuity from one seismic trace to the next. Coherent seismic noise trains however, do persist in organized form from trace to trace yet again convey no direct geologic information.

Source-generated noise in the form of surface waves exhibiting strong coherence and obscuring all reflection events is illustrated by the exhibit taken in part from Jolly and Mifsud. The immediately following exhibit provided by Western Geophysical Company of America shows yet another problem - multiple reflection events. In this case, the air-water interface and the water-water bottom interface allow a sequence of strong reverberations to occur. These appear with regularity at successively later times and can obscure the simultaneously arriving primary reflection events which we wish to interpret.

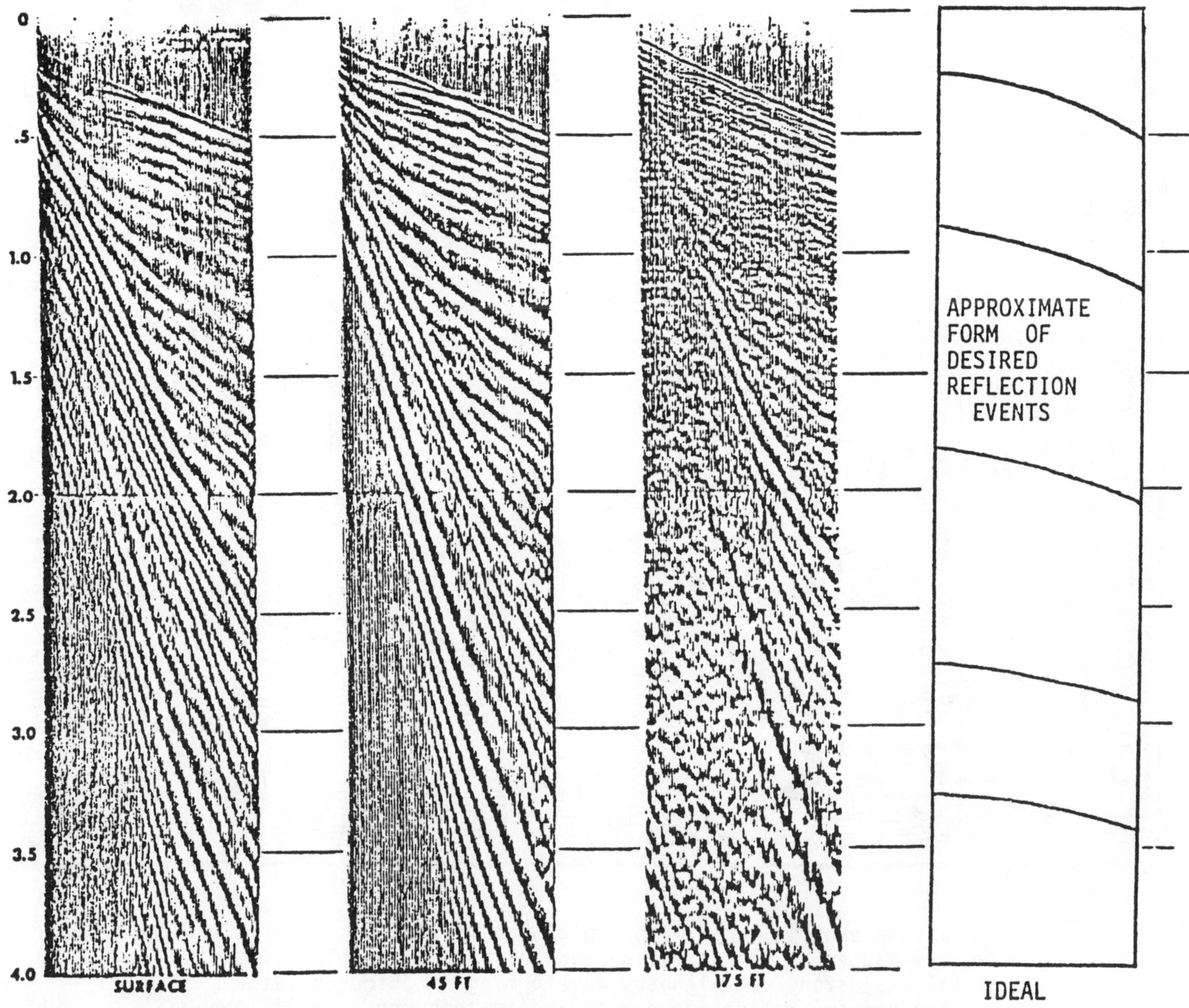

SURFACE WAVES FOR DIFFERING SOURCE DEPTHS (AFTER JOLLY & MIFSUD)

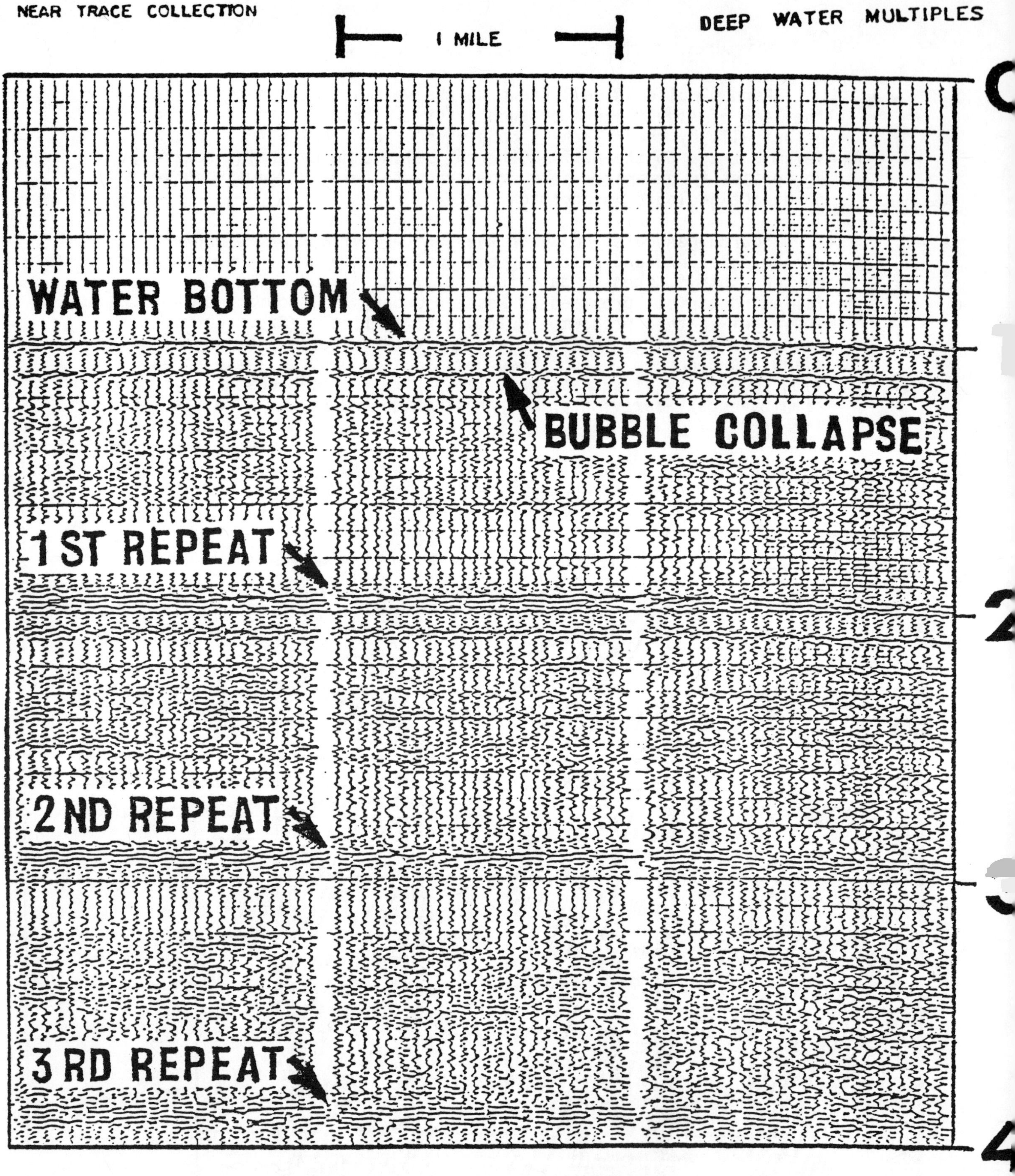

In the exhibit, apart from multiples we see also a very lengthy waveform caused by the formation and collapse of bubbles after the initial explosion. The lengthy waveform again obscures later arriving primary reflection events and in this sense limits the resolution of our interpretation.

Clearly then in making our seismic interpretations and correlations for stratigraphic objectives, we must make certain that the events we are treating do in fact correspond to acoustic boundaries. We have just seen examples of two kinds of common seismic events not of this nature. The major objective of most of the processing to which seismic data is subjected is to present the interpreter with events of the type he desires - namely those which correspond to changes in acoustic parameters. Processing is rarely accomplished perfectly however and we must admit at all times the possibility of observing coherent or incoherent seismic events which are not simply related to acoustic impedence contrasts.

The matter of departures from a flat but continuous and homogeneous subsurface introduces other considerations of great interpretive importance, certain of which may be quite subtle. At the more obvious level we note from the two simple ray trace models of the next figure and the resulting time sections that in the presence of structure, seismic events do not appear at trace locations corresponding to the locations of the reflection point in the subsurface. As a corrolary, it is clear that anticlines will appear greater in extent of their seismic expression than they are in fact. Analogously, synclines exhibit reduced seismic extent. In the model having a dipping pinchout, note that the reflections appertaining to the possible prospect at the termination can only be observed over acreage which has not been leased.

Diffraction events from the pinchout are also shown. These are yet another type of seismic event which although coherent in character do not simply portray material boundaries. Rather, diffractions are related to abrupt discontinuities, inhomogeneities or tight curvature and usually have hyperbolic form. In the presence of severe sturcture, they can account for the majority of the seismic events to be noted. This is illustrated by the schematic overthrust fault model study shown in the next figure along with its wave theory seismic response. The hazards of stratigraphic interpretations in such a province are all too evident. For this illustration, the reflection events corresponding to the hydrocarbon-water contact appear only on traces 23, 24, 25, 26 as a slightly dipping horizon at 1.5 sec.

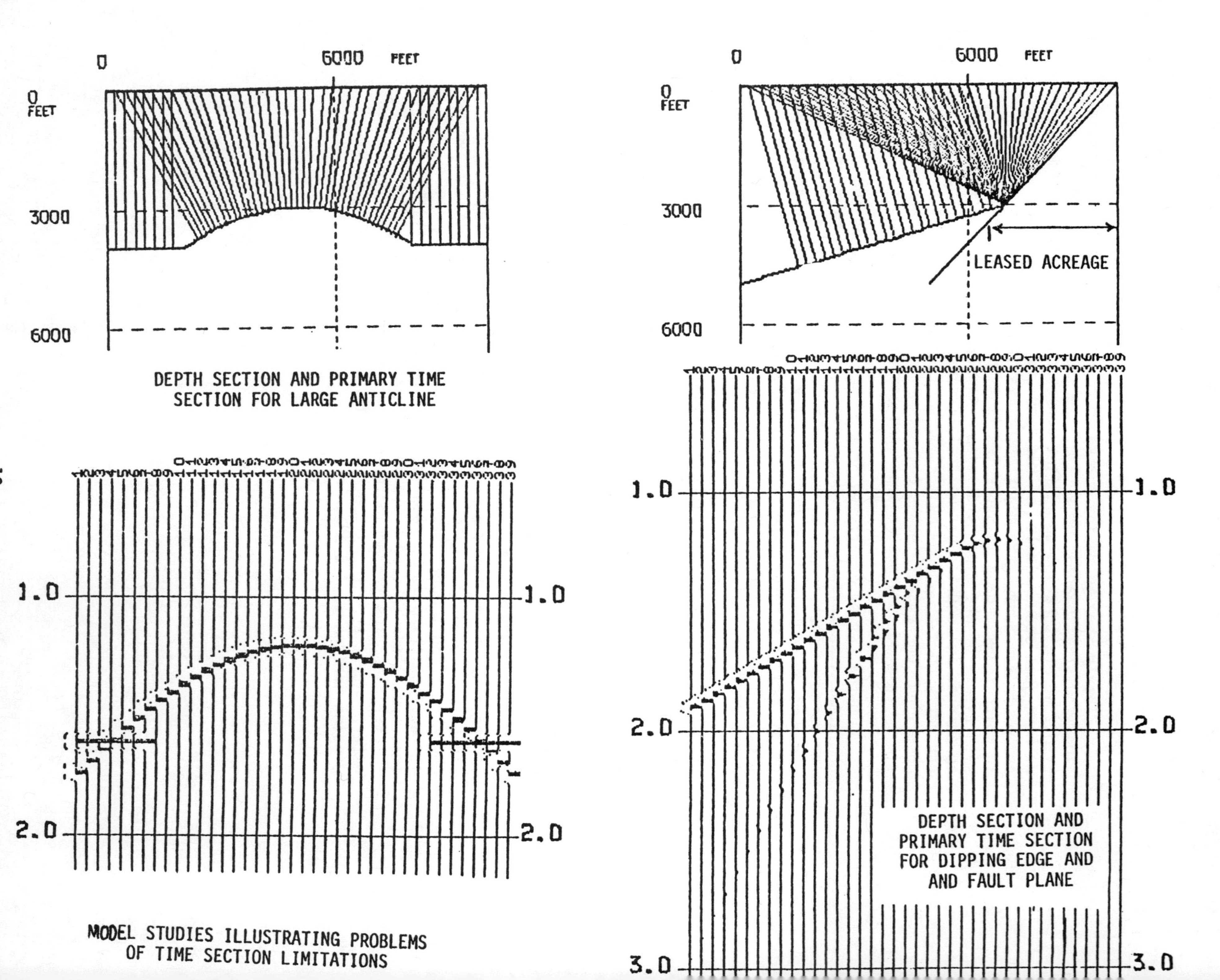

DEPTH SECTION AND PRIMARY TIME SECTION FOR LARGE ANTICLINE

DEPTH SECTION AND PRIMARY TIME SECTION FOR DIPPING EDGE AND AND FAULT PLANE

MODEL STUDIES ILLUSTRATING PROBLEMS OF TIME SECTION LIMITATIONS

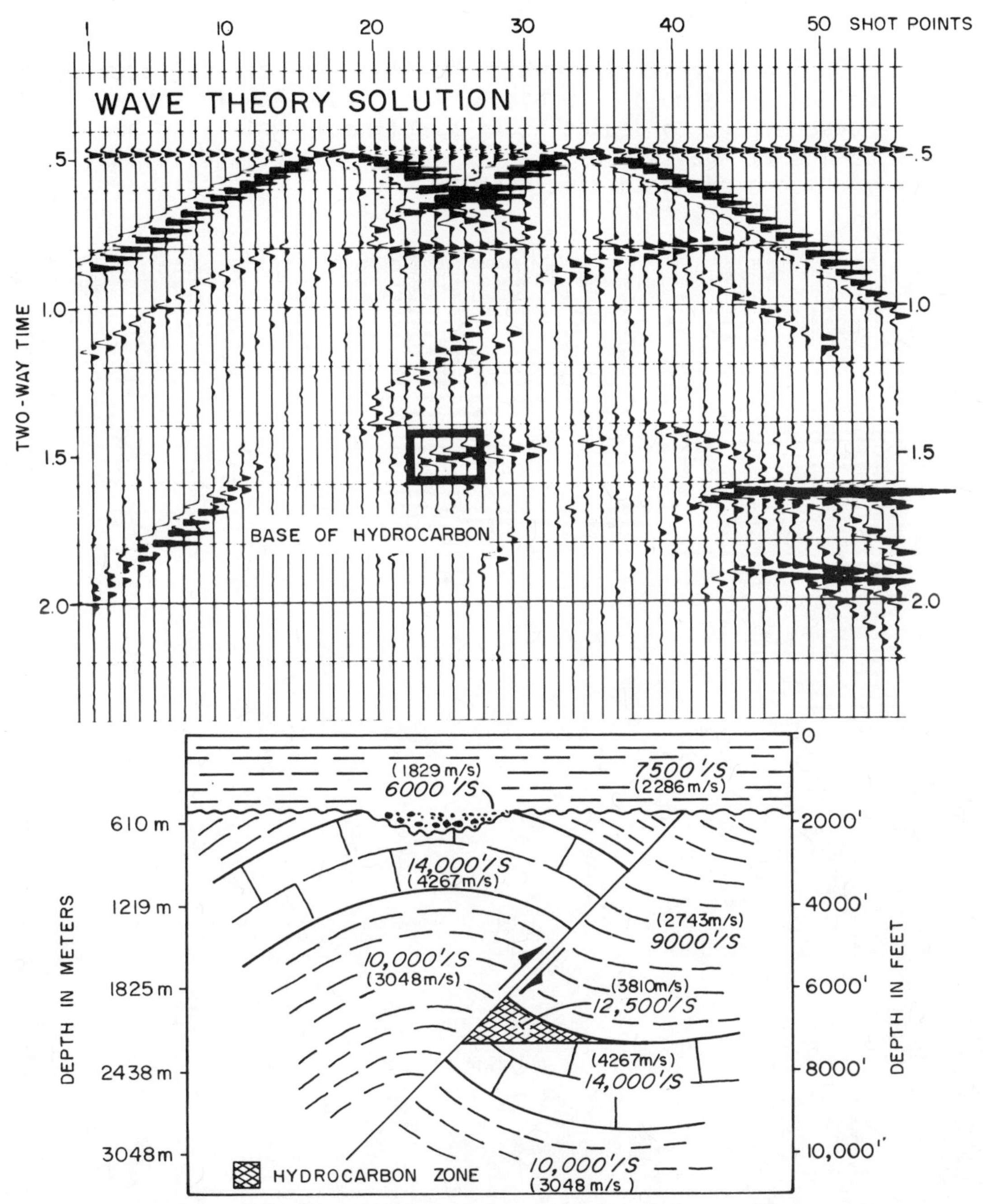

—Overthrust-fault model and computed seismic expression. Diffraction events dominate seismic picture, and bending of seismic rays obscures identification of reflections from hydrocarbon zone.

Seismic data processing again offers attempts at presenting to the interpreter data more closely corresponding to ideal circumstances. Migration of seismic sections seeks to portray the subsurface with reflection events corresponding to their points of origin. In migrated sections, anticlines and synclines should appear in their true dimension and tight curvatures and bed terminations portrayed accurately without "masking" or "smearing" caused by diffractions. Again, processing is not often perfectly accomplished and the interpreter is forced to look beyond these obstacles in seeking to define stratigraphy.

The pair of sections which follow have been migrated using some techniques developed by Digicon, Inc. Note how the "bow tie" signatures are unscrambled to depict tight synclines. Note also how the large anticline at the left of the section is compressed to its true dimension on the migrated depth-time display.

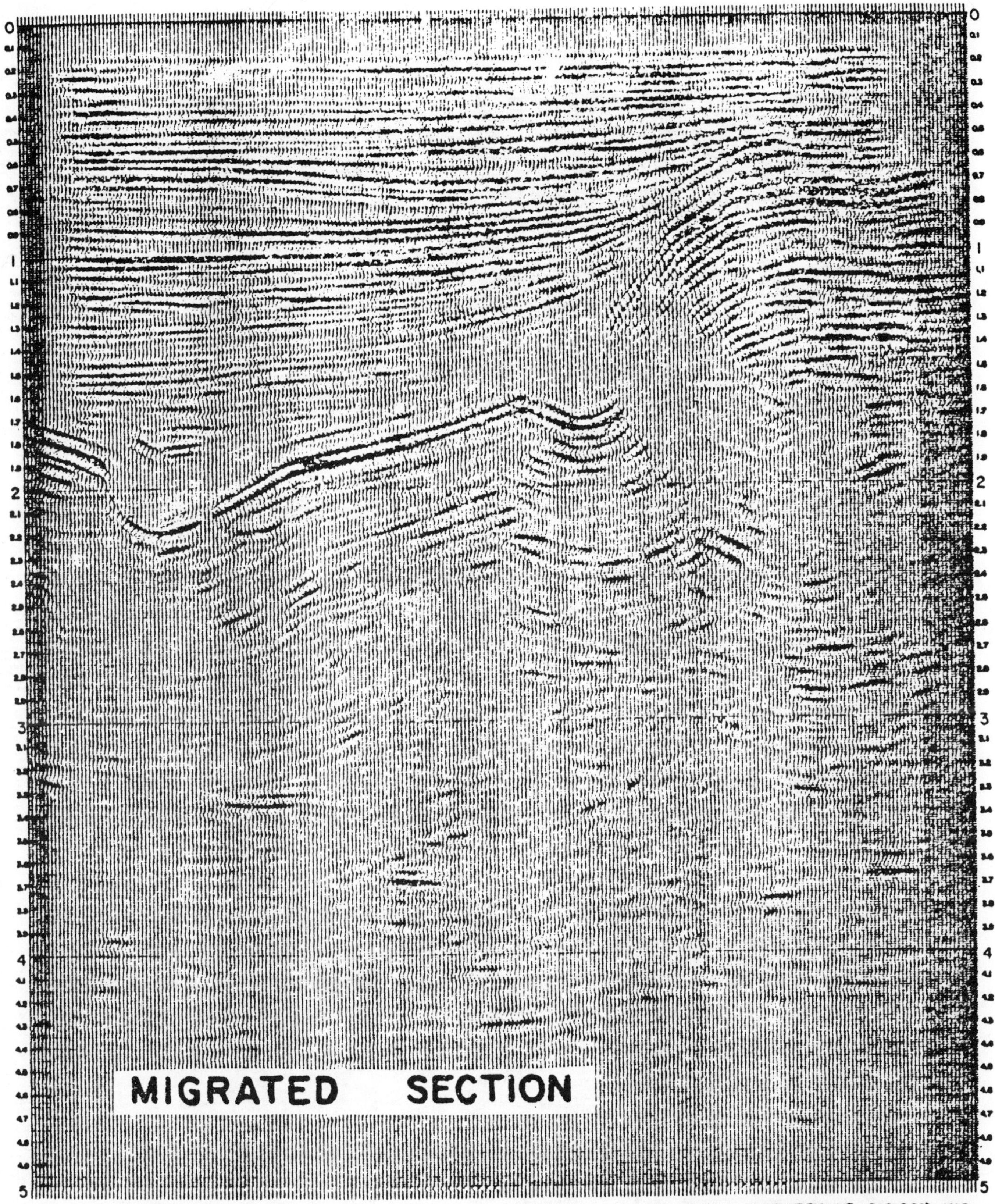

Hence, migration seems to take us from what at first seems an insurmountable problem again to a state where the ideal world interpretive method may be applied. Unfortunately, as we cautioned earlier, we cannot always count on the processing techniques to work perfectly, if at all.

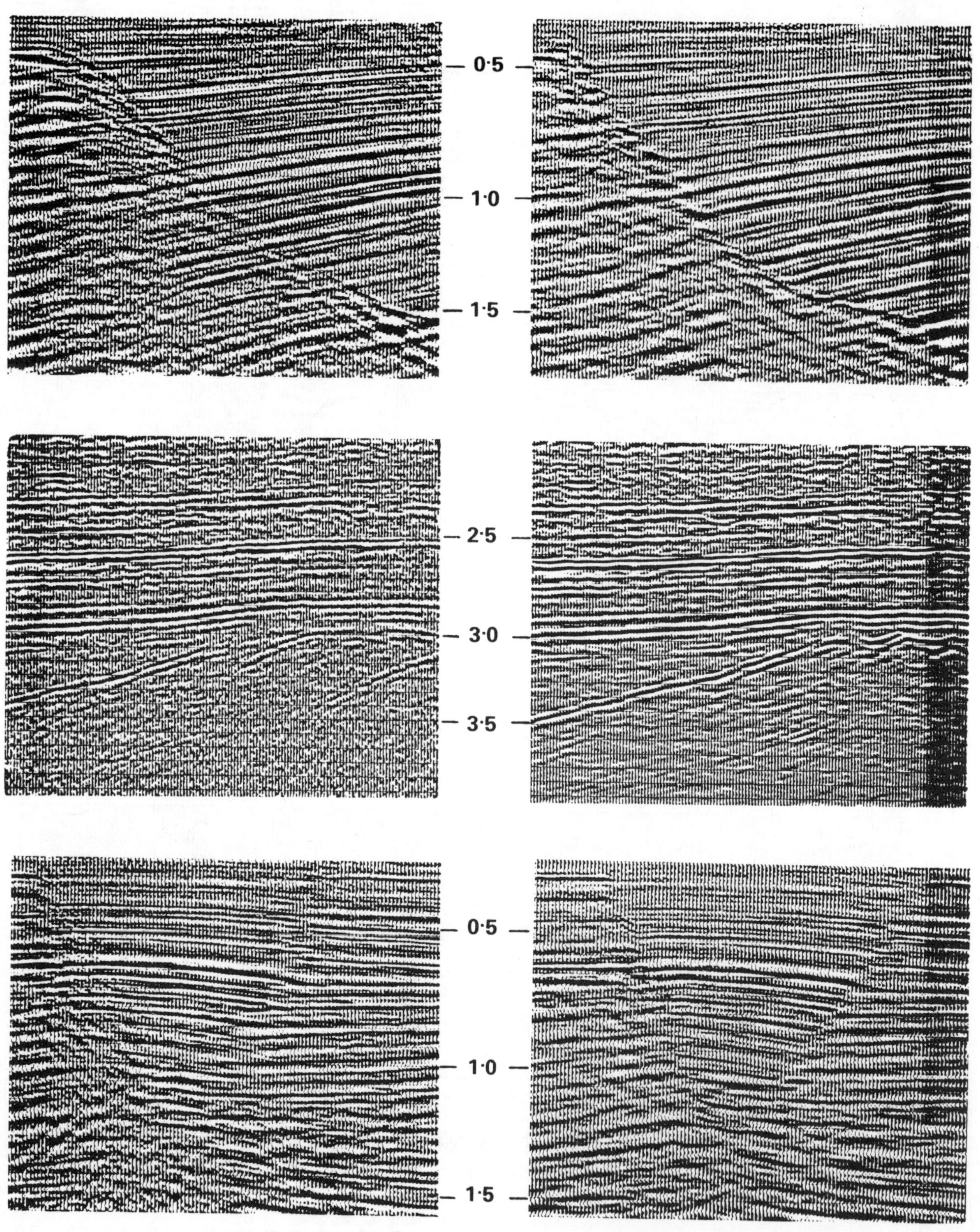

Before Migration

After Migration

The facing data panels representing "before" and "after" migration illustrations developed by S & A Geophysical leave little doubt as to the benefits derived by migration. This is, of course, essential in a stratigraphic context. It is particularly interesting to note that even in cases where migration fails to perform the task we ask, the result tends toward being correct although it falls short of the goal. In the illustrations, the fault, angular unconformity and Horst block are all considerably improved in appearance.

We should comment also returning to note the previous example - pair by Digicon, that migration may in fact be a prerequisite to application of the stratigraphic methods described by Vail and the various EXXON authors in the AAPG Stratigraphic Seismic Monograph Number 26. Clearly, if inferences are to be made from the attitude and relative positioning of reflectors, these must be corrected first to present a more realistic picture of the subsurface.

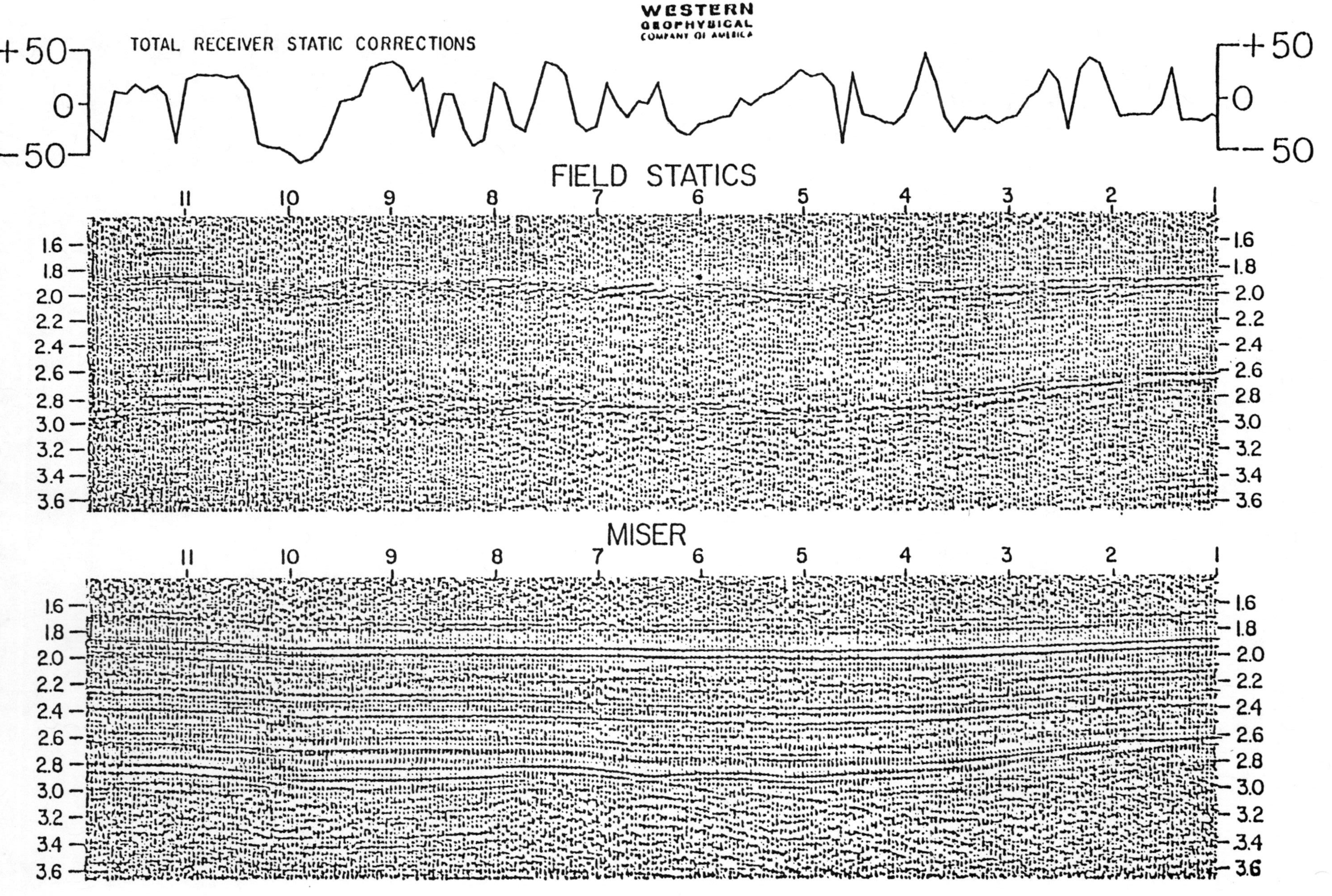
WESTERN
GEOPHYSICAL
COMPANY OF AMERICA
TOTAL RECEIVER STATIC CORRECTIONS
+50
0
-50
FIELD STATICS
MISER
11
10
9
8
7
6
5
4
3
2
1
1.6
1.8
2.0
2.2
2.4
2.6
2.8
3.0
3.2
3.4
3.6

At a more subtle level, departures from linearity or small inhomogeneities at the near surface can cause great disruptions in the appearance of seismic sections, rendering the result uninterpretable. The cause of the problem rests with the introduction of trace-to-trace time shifts (static time shifts or delays). These cause the usual processing sequence, which attempts to attenuate noise, to fail. An illustration of such a problem, along with a processing technique to eliminate the problem is also shown. The particular process MISER, by Western Geophysical Company of America, also gives an indication of the trace-to-trace time shifts which had to be introduced to treat the difficulty. It proved necessary to include time shifts having magnitudes of + and -50 msec., as referred to a datum as corrections at various receiver locations for this data set.

Note that if the interpretive task to be addressed consisted only of mapping the two major reflective horizons, no corrective processing would have been needed. On the other hand, if we wish to detect reefing just above the deeper horizon or else some similar stratigraphic objective, a technique as illustrated is a vital step.

The pair of sections which follow note how a reef in Kalkaska County, Michigan can be recognized only after the application of a statics correction. Glaciation has caused the near surface problem in times past. The particular example is taken from McClintock (1975) who notes also that amplitudes of the seismic data also play a role in the detection of reefs. We shall describe the importance of seismic amplitudes also in the text which follows.

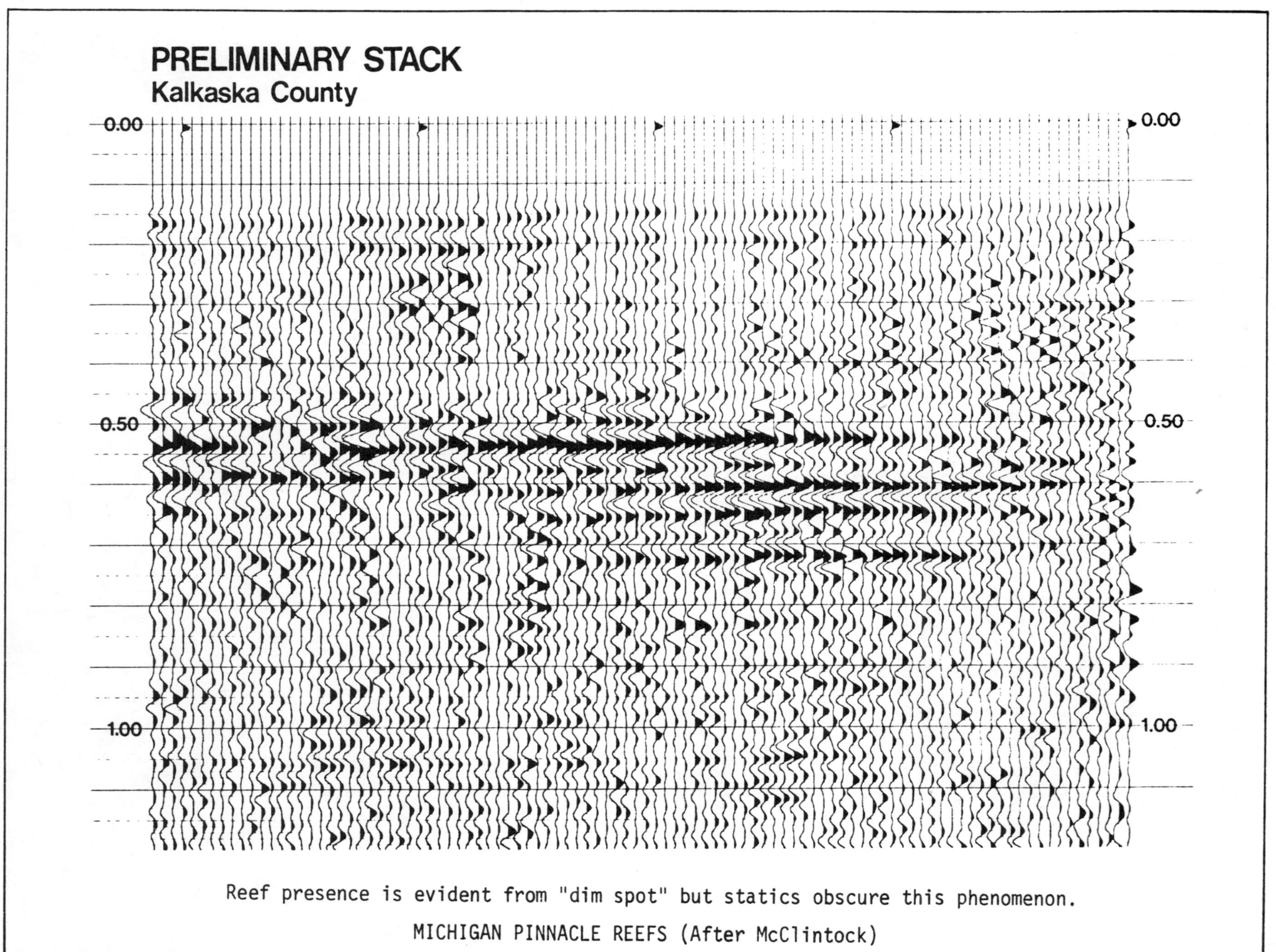

Reef presence is evident from "dim spot" but statics obscure this phenomenon.

MICHIGAN PINNACLE REEFS (After McClintock)

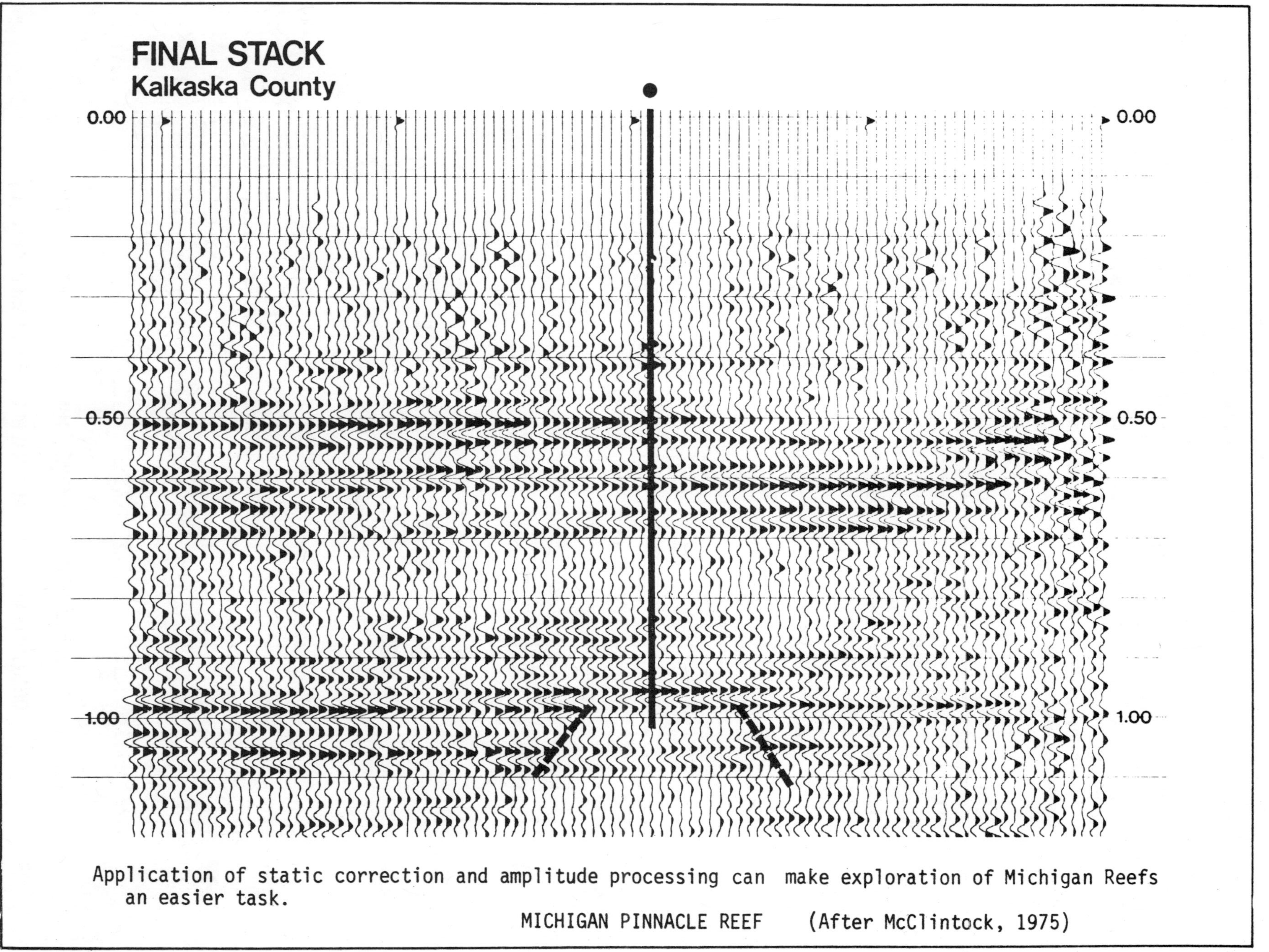

Application of static correction and amplitude processing can make exploration of Michigan Reefs an easier task.

MICHIGAN PINNACLE REEF (After McClintock, 1975)

The question of the seismic waveform and the stratigraphic resolution ultimately obtainable using any particular waveform requires a discussion at some length. At this time we shall simply note a comparison of similarly processed time sections over a North Sea oil field. In one panel the original waveform is present while in the second, a shorter waveform of simple symmetric character has been introduced. An increase in resolution of the reflection defining the structure is most evident. No direct comparison of the two sections is possible owing to the differences in waveform. Here again we must emphasize that significant correlations of seismic results with the geologic inputs requires a strong correspondence of waveforms.

Of course, to attempt stratigraphic interpretation from seismic data we must expect to have good-quality data which has been acceptably processed. Apart from consideration of the factors already described, we must further be prepared to accept the subsurface as three-dimensional (instead of two-dimensional as implied by seismic profiles) and treat also both lateral and vertical transitional effects which are known to occur in nature as an intimate part of stratigraphic mechanisms.

A. PREDICTIVE DECONVOLUTION BEFORE STACK

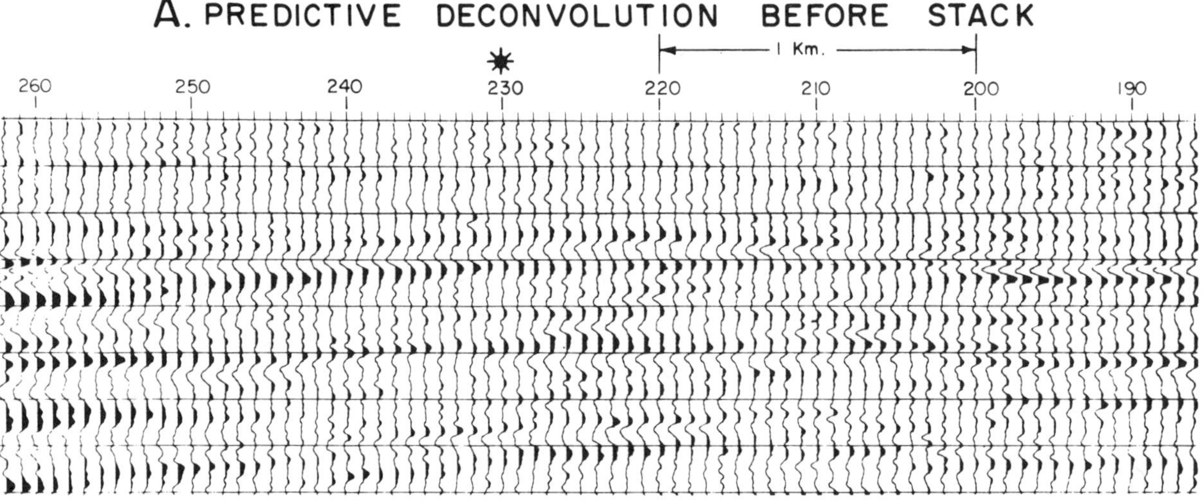

B. WAVELET PROCESSING BEFORE STACK
PREDICTIVE DECONVOLUTION AFTER STACK

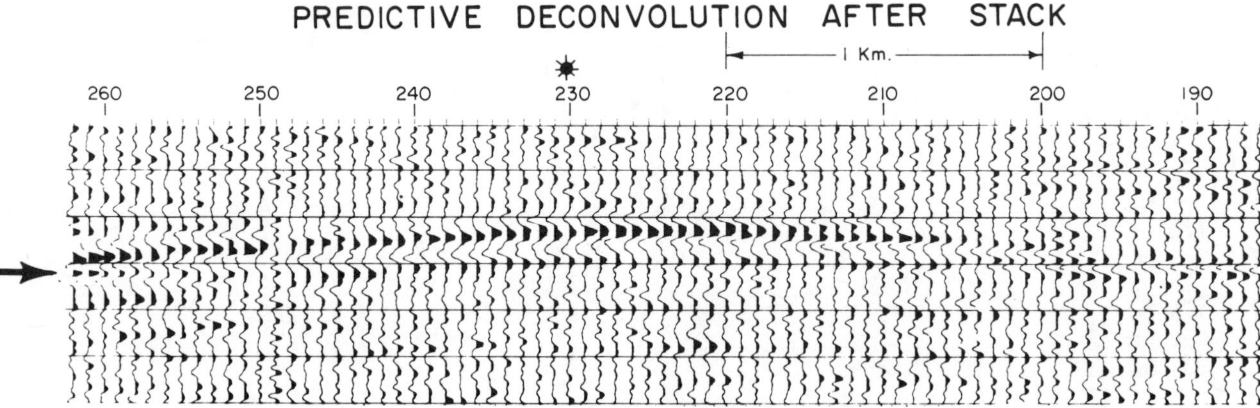

PROCESSED SEISMIC EXPRESSIONS OF OIL FIELD WITH

ORIGINAL WAVEFORM AND WAVELET PROCESSED RESULTS

Let us set forth an interpretation procedure for handling seismic data which begins with a processed seismic section. For the moment we will suspend all questions or doubts about the effectiveness of processing and interpret the data based on the following assumptions:

> Each trace of the section represents only primary reflections from the subsurface having location immediately below where we have shown them to be plotted.
>
> Individual reflection events can be identified and their amplitude is diagnostic of the step in acoustic impedance across the boundary causing the reflection.

In this ideal world then, we may very simply state the objectives and procedures of both seismic and stratigraphic interpretation. The accompanying figure from Dedman, Lindsey and Schramm (1975) shows a portion of a lithology log and a corresponding acoustic impedance log. Each contrast in acoustic impedance will be marked by a reflection event having a simple waveform. The polarity or sense of the reflection and its size will indicate the nature of the contrast. Individual reflection events for the model are shown along with their superposition in the resulting seismic trace.

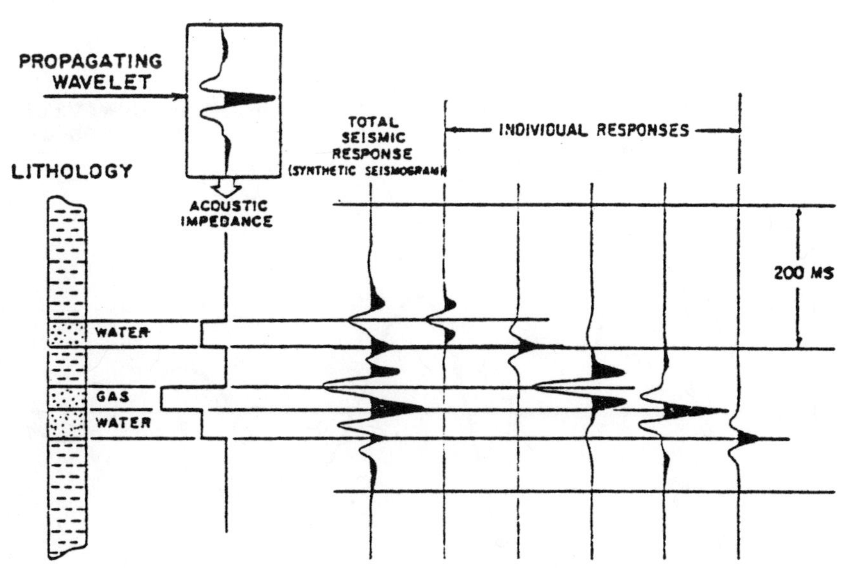

—Relationship between lithology, propagating wavelet and seismic response.

The question of the shape of the seismic waveform next must be addressed if stratigraphic interpretaion is to be considered. We have already seen in the case of a bubble pulse sequence how much of a problem an unfortunate seismic pulse length, or for that matter pulse shape, can be. We can reenforce the significance of this statement by considering on the following figure our ideal subsurface, but now viewed through a more typical seismic pulse.

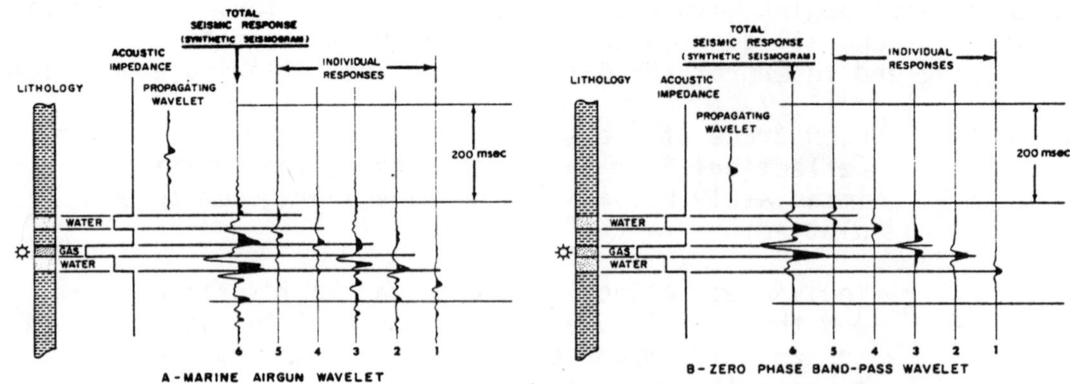

If we wish to recognize separate reflections and talk of them in terms of acoustic impedance steps, we are at a decided disadvantage. Hence, some corrective step in processing is clearly needed to condition the waveform to a shape providing greater interpretive leverage.

Problems associated with a complex waveform shape have long been recognized and, in fact, a certain class of the deconvolution methods known as "spiking" deconvolutions had been devised earlier to treat the problem. As a result of such a study, in 1957 Dr. E. A. Robinson produced one of the earliest estimates of a seismic waveform, which is shown next.

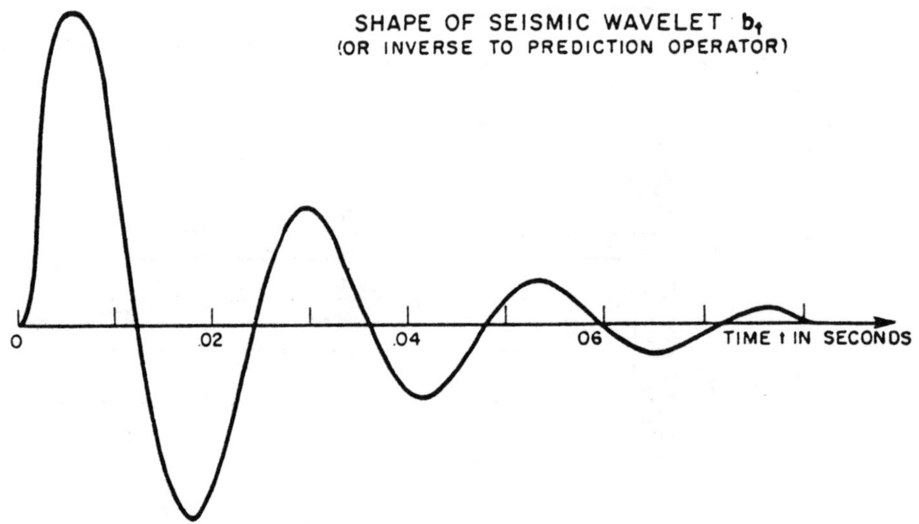

More recently we have recognized that seismic wavelets can be estimated or measured, and that transformations of shape based on such estimates are more likely to be correct than purely statistically conceived approaches like deconvolution. Methods of this type are collectively known as wavelet shaping or wavelet processing, and must be discussed more fully.

Stratigraphic interpretation would begin then with the development from each trace of an acoustic impedance log or equivalently a reflectivity series. These results would be correlated from trace-to-trace to provide the structural considerations and correlated also with whatever geologic information is available. Using geological principles and insights as appropriate to the region lithologic estimates would be inferred, and from these estimates and the indicated changes, geometry and depositicnal patterns, sequences and history might be interpreted.

The departures from the ideal however, have been previously emphasized and we see that our objective may not usually be directly accomplished. Nevertheless, those tools which can assist us in our goal may now be identified and the importance of their specific roles plainly appreciated.

Good data acquisition and processing are first taken as necessary prerequisites for any seismic stratigraphic work. It is further obvious that some ability to capture, manipulate and generally transform seismic waveforms is needed in order to facilitate the correlations which must be made. Such tools and techniques and their effects on waveform polarities and amplitudes are described in the next sections. We also understand that geometric effects, diffractions and other complicating factors and elements must be included in our interpretations. The role of modeling and in particular two and three dimensional wave theory modeling will be discussed specifically in this regard. Finally, the question of developing principles for making lithologic estimates, improving the vehicles for the correlations and quantifying the resolution inherent in the seismic expressions must be addressed.

II. INFORMATION CONTENT AND RESOLUTION POTENTIAL OF SEISMIC DATA.

A. Seismic Data as The Propagation of Waves

A question never answered satisfactorily to this day is what precisely can seismic data tell us about the subsurface? Subsidiary aspects of this same question might be: How small a feature in the subsurface can the reflection seismic method resolve?; and How does seismic data relate to high frequency acoustic data acquired in boreholes? Addressing these questions requires that we first adopt a wave theory viewpoint toward the propagation of a seismic disturbance.

Let us first resolve a paradox which appears to be presented on the next figure. If we consider a colocated source and receiver, then according to familiar ray theory we would draw a raypath down to the reflecting boundary and back. We would call this "the reflection" and the point of contact on the reflector "the reflection point".

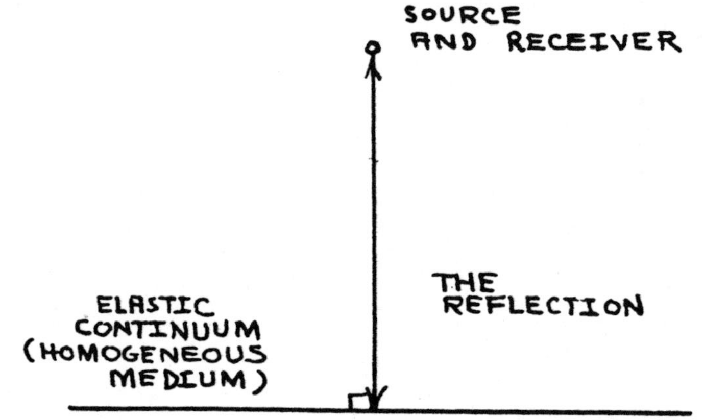

At this point we shall redraw this same picture with Huygen's principle of wave theory in mind. Our task will then be one of reconciliation.

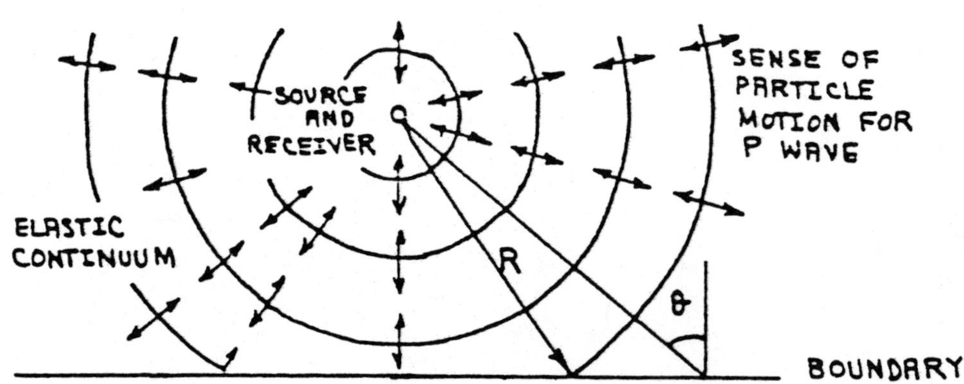

In the preceding figure we see that in time the propagation wavefront will strike the entire boundary and not just a single point. According to Huygens's principle then, every point of the boundary will act as a source and reradiate energy to the receiver. Hence in wave theory, unlike ray theory, every part of every boundary contributes to the seismogram received anywhere for any source position.

If we are to reconcile the ray theory view with the wave theory picture, then we must inspect the situation more closely. It should first be obvious that for a given source location, each point on the boundary does not make the same contribution to the receiver. Specifically then, the magnitude of the contribution for each part of the boundary depends on two factors:

R - the distance from the source to the part of the boundary (to account for divergence effects).

θ - The inclination of the particle motion (measured normal to the boundary).

When $\Theta = 0^{o}$ for any fixed value of R, the returned energy is a maximum, whereas when $\Theta = \pm 90^{o}$, we will have the minimum contribution. This is intuitively reasonable since for $\Theta = \pm 90^{o}$, the particle motion for the case shown will be propagation in a sense parallel to the boundary and so not directing any disturbance toward the receiver.

Should a boundary have a geometry which is quite similar to the shape of the wavefront, then all parts will tend to contribute strongly giving a strong or focused reflection. This effect is quite well known as it is also responsible for the dimming of reflection amplitudes which follows from anticlinal reflectors.

With these concepts in mind, the effective zone which can be said to be responsible for the seismic reflection can be readily defined. This zone is termed the Fresnel Zone.

The figure which is next encounted illustrates the Fresnel zone.

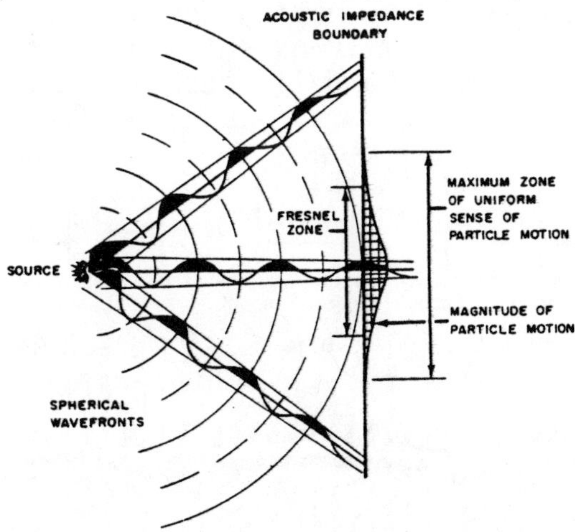

—Spherical wave motion encountering a flat reflector illustrating Fresnel zone region which is principally responsible for reflection event.

Spherically radiating wave motion encounters a flat acoustic impedance boundary. The maximum-sized region of this boundary, over which the particle motion is all in the same direction, is indicated. A Fresnel zone, although directly related to the size of this region, has a somewhat smaller value since some allowance must also be made for the respective magnitudes of particle motion within this region.

If the source is at a distance R from the reflecting boundary where R represents travel distance at a velocity V over a two-way time of t seconds, and the dominant frequency content of the seismic data is f_c, then a Fresnel zone radius r_f is approximated by

$$r_f \simeq \frac{V}{4}\sqrt{\frac{t}{f_c}}$$

This mathematical relation is derived using basic geometry and an emperical weighting. For the values V = 3,000 m/sec, t = 1 sec, and f_c = 25 Hz, r_f = 150 m.

Seismic illimination of the subsurface may then be considered analogous to a "searchlight beam", and the subsurface area illuminated by this beam is considered to be the Fresnel zone. With such an intuitive point of view, we can consider fundamental questions such as how small in extent a subsurface feature must be as compared to a Fresnel zone in order to be detectable, and what the nature of its seismic expression must be in order to facilitate detection.

B. Resolution Capabilities of Seismic Data

In the next figure note that, as a sandstone body approaches the lateral dimension of a Fresnel zone and becomes still smaller, the seismic signature loses all reflection character and appears in the diffraction form corresponding to a reflecting point. The lessons are clear: (1) Thin beds of small spatial extent have a seismic expression, but not dominantly of reflective nature; and (2) although the form of the signatures are alike for sandstone bodies smaller than the Fresnel zone, their size can be gauged by noting the differences in the strengths of the seismic events. The smaller extent bodies produce weaker events by occupying only portions of the full Fresnel zone.

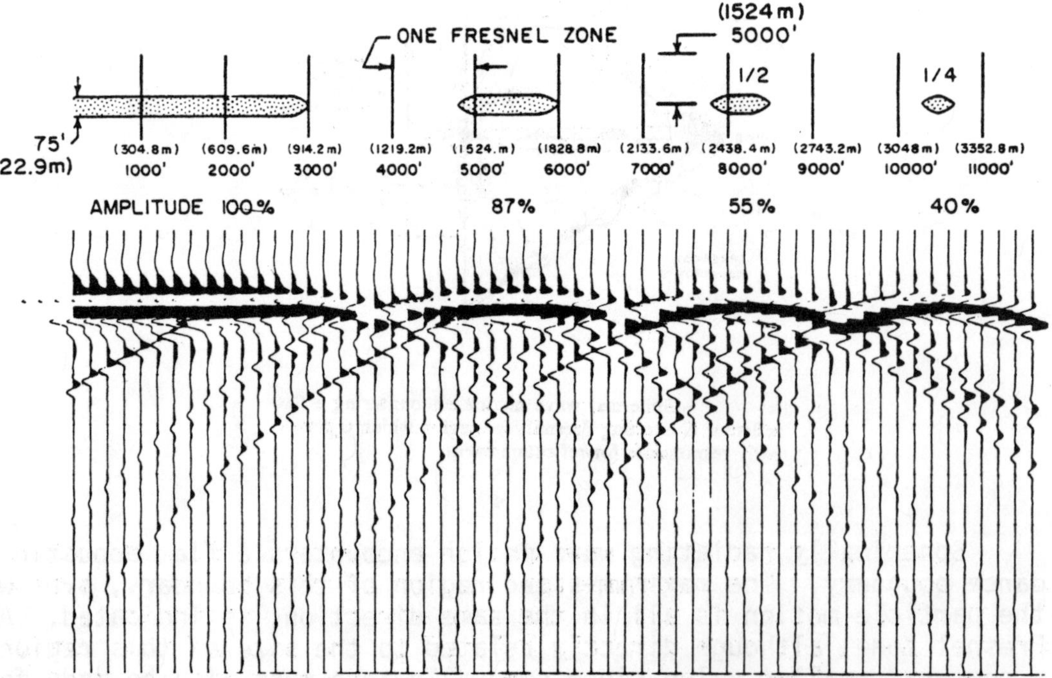

-Wave-theory model response for sandstone bodies of varying lateral extents illustrating significance of Fresnel zone size.

Amplitude processing is concerned with the preservation of seismic strengths. With the mechanism just observed it was seen that, where relative amplitudes have significance, they may be related to the identification of small lithologic inhomogeneities.

As a practical matter, the size of the Fresnel zone is initially related to the quarter wavelength of the dominant frequency component of the seismic waveform and the near-surface velocity. For the expression for r_f, the size of the Fresnel zone grows in proportion to the stacking velocity which may be determined readily from seismic data. The case study examined by the figure is a typical one for intermediate-depth Gulf Coast sandstones.

Next we can investigate the role of the Fresnel zone from the point of view of a three-dimensional wave-theory model. Our figure shows a reef-like box structure over which two seismic profiles are calculated. Line 1 crosses the structure through its long axis; Line 4 parallels Line 1, but is 152.4 m from the edge of the structure. The structure is 1,542 m below the surface.

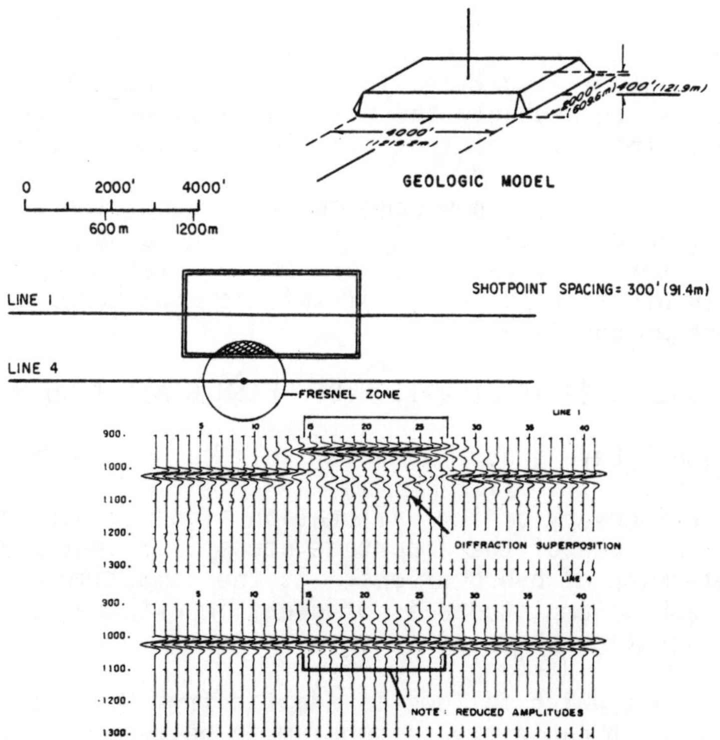

—Three-dimensional wave-theory modeling over "box car" reef illustrating significance of Fresnel zone in producing reflections.

Line 1 shows the structure clearly, along with all diffraction events at its termination. The superpositon of the diffraction events from the horizon at the base of the reef almost suggest the continuity of this horizon under the structure. Observe that reflection amplitudes over the reef are comparable in magnitude with those of the reflective horizon on which it was developed.

Line 4 suggests a structure above and below (though more subtle here) the reference horizon. The reflections above the reference horizon are from the flat-topped portion of the structure. The Fresnel zone which gives rise to reflection events lies partially on the reef top and partially off of it. In fact, the reduced-amplitude event seen for the reef top is proportional to that part of the Fresnel zone area which is on top of the reef. The reference-horizon reflection amplitude seen below the reef-top reflection is also reduced since it, too, represents only a part of the Fresnel zone (although this effect is more difficult to see). Below the reference horizon the weak events are formed by re-enforcement of diffractions from the base edges of the reef acting out of the profile plane.

Three-dimensional wave theory studies show that all seismic interpretations must be visualized in a three-dimensional environment. Such model studies are probably the most powerful means presently available for teaching the principles of seismic interpretation.

Having established now some concepts about the information content and resolution potential of seismic data, we should now consider several of the more currently developed processing steps which seek to help us use such information. The use of seismic amplitudes is the first of these which we consider.

III. PROCESSING FOR PRESERVATION OF SEISMIC AMPLITUDES

A. Signal Levels and The Seismic Trace

The rediscovery of the exploration significance of seismic event amplitudes was one of the important steps in recent progress toward direct detection of hydrocarbons. At the same time we appreciate that a closer look at amplitudes is of great importance, particularly in search of stratigraphic objectives.

The figure which follows is adapted from Lindseth and notes the mechanisms which cause the seismogram to be generally weaker with increased reflection time. Against this background the amplitude distinction between a strong reflection and weak one as shown is a secondary matter.

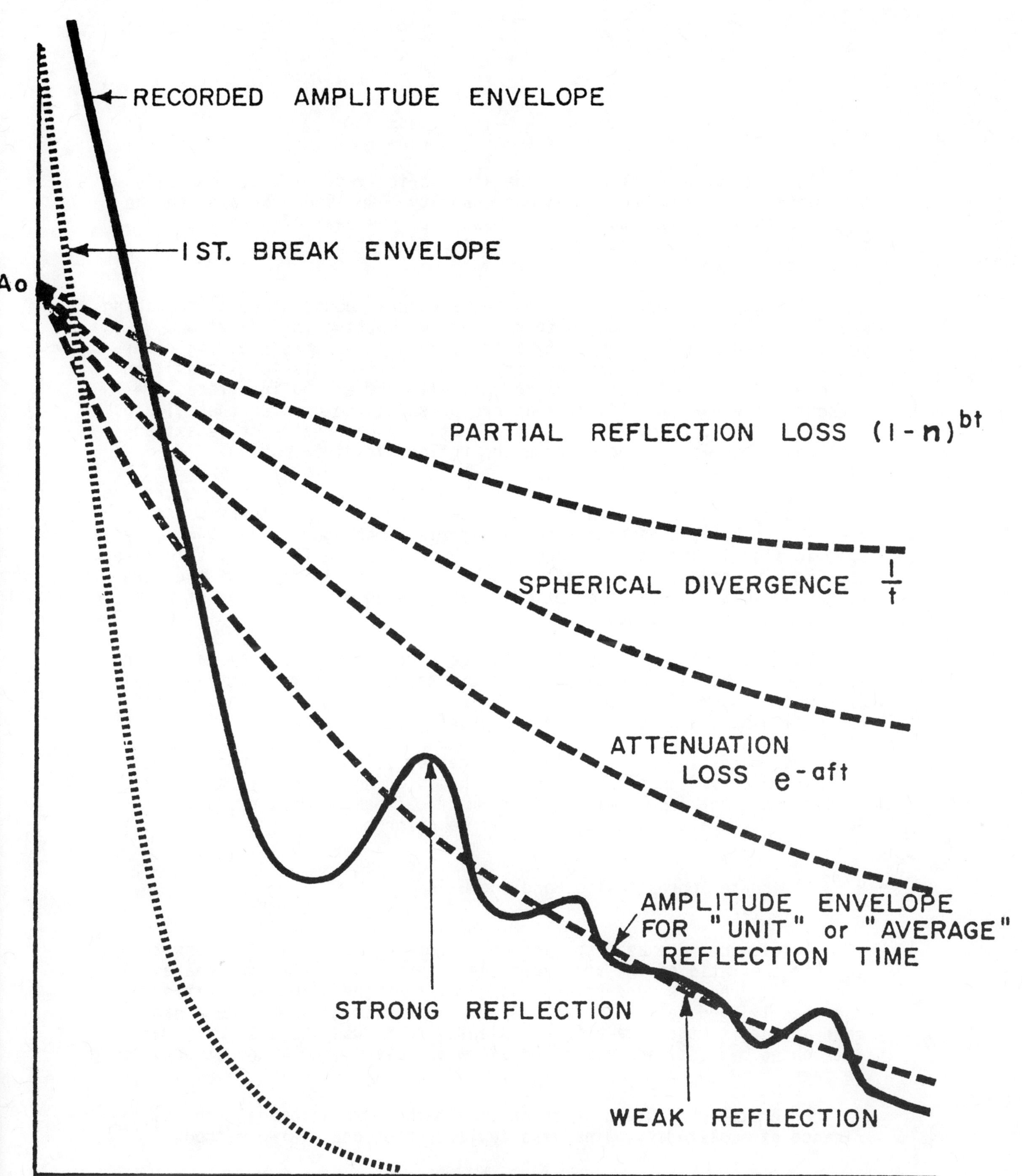
RECORDED AMPLITUDE ENVELOPE
1ST. BREAK ENVELOPE
Ao
PARTIAL REFLECTION LOSS $(1-n)^{bt}$
SPHERICAL DIVERGENCE $\frac{1}{t}$
ATTENUATION
LOSS e^{-aft}
AMPLITUDE ENVELOPE
FOR "UNIT" or "AVERAGE"
REFLECTION TIME
STRONG REFLECTION
WEAK REFLECTION
TIME

In fact, we are not really speaking strictly of reflection amplitude which is a frequency dependent quantity. Amplitude as used in the present context refers to a gross measure of the peak of a reflection envelope and more correctly should be termed "reflection strength" or "reflection level".

Divergence losses are, of course, a natural consequence of spreading wavefronts and would occur whether or not reflecting interfaces or velocity gradients were present, though these govern the specific divergence behavior. Partial reflection losses (or transmission losses), on the other hand, occur only because partially reflecting interfaces are present. The more interfaces that are present, the greater these losses will be. Hence, transmission loss effects can be greater or less than divergence losses, depending upon the reflective distribution.

Attenuation losses are least well understood of all these mechanisms. They are bound up both with the absorptive process which turns seismic energy into heat, and scattering by material inhomogeneities which makes a seismic signal lose its coherence. Both effects are frequency selective causing high frequency components to be lost most rapidly. Of the loss mechanisms cited, only attenuation is associated with a change of waveform, which of course has interpretive significance.

Loss mechanisms as described can be highly variable from one ground location to the next and cannot be estimated with any accuracy a priori. Even when information from well logs is available, the general loss of reflection strength is often not successfully predicted.

Treatments to preserve significance for seismic amplitudes then, whether conceived on some theoretical basis or not, usually resort also to empirical methods or else some form of calibrated measurement. Several of the alternate methods will be described.

B. Alternate Approaches to Handling Gain

A seismic trace as displayed on a section has a value of less than 100 for the ratio of the strongest event level to that of the weakest discernable one. We understand, however, that the comparable ratio in the underlying data may in fact be 10,000 or even greater. Gain has been applied to the data prior to display to reduce the size differences introduced by loss mechanisms and allow all events, weak and strong, to be seen.

Prior to development of processing methods to accomodate the significance of amplitudes, gains were applied using one of two methods:

Adaptive gains

Programmed gains

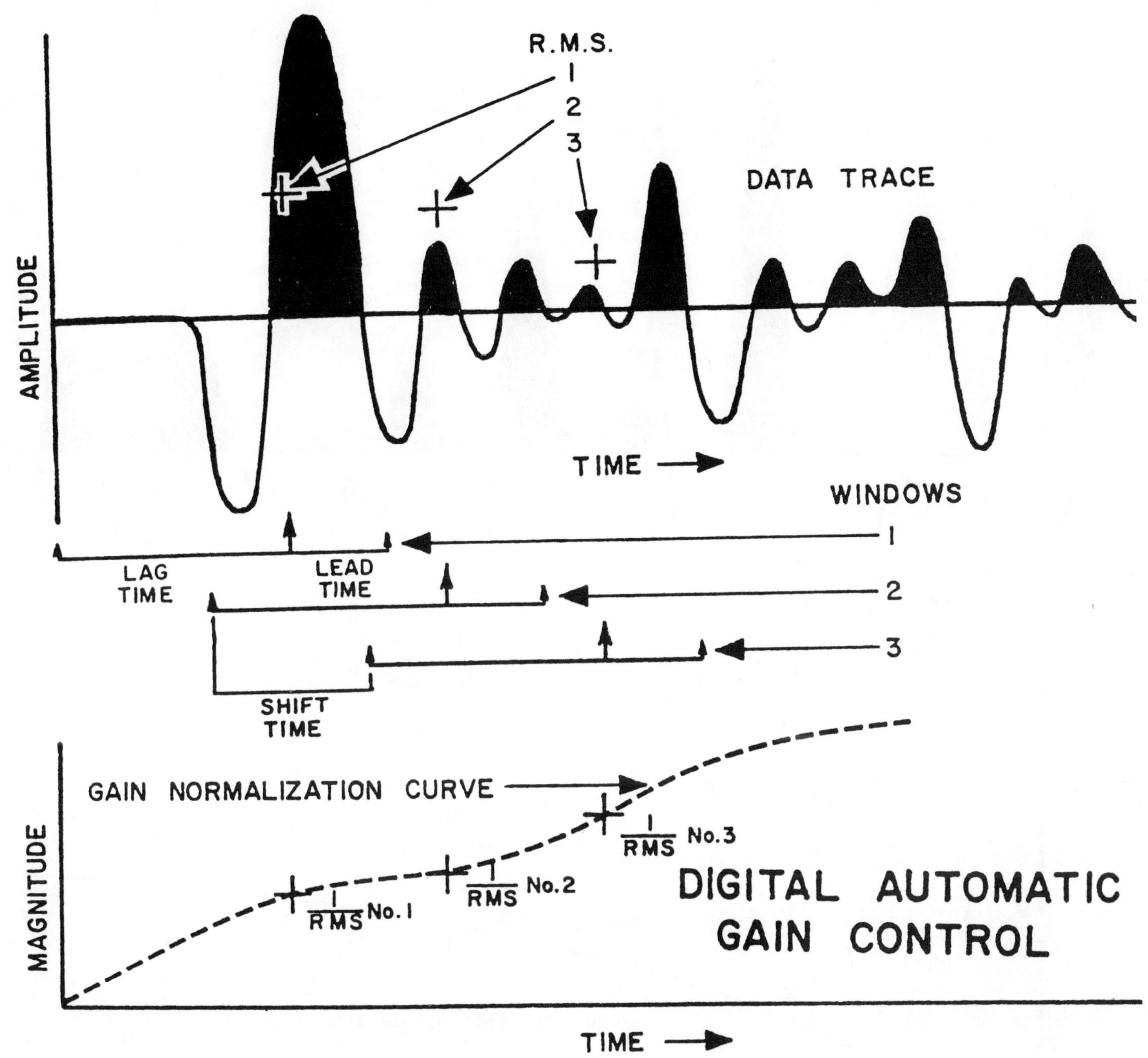

Adaptive gains were derived from running estimates of the trace signal level made over "windows" of prespecified size and overlap. Using these values, amplification factors were computed for application within the windows as illustrated in the following figure from Lindseth. Such gains were typically called Automatic Gain Controls or "AGC". It is important to note that unless the amplification factor is applied at the center of the window of its derivation, a signal phase distortion (or equivalently alteration in shape) can occur.

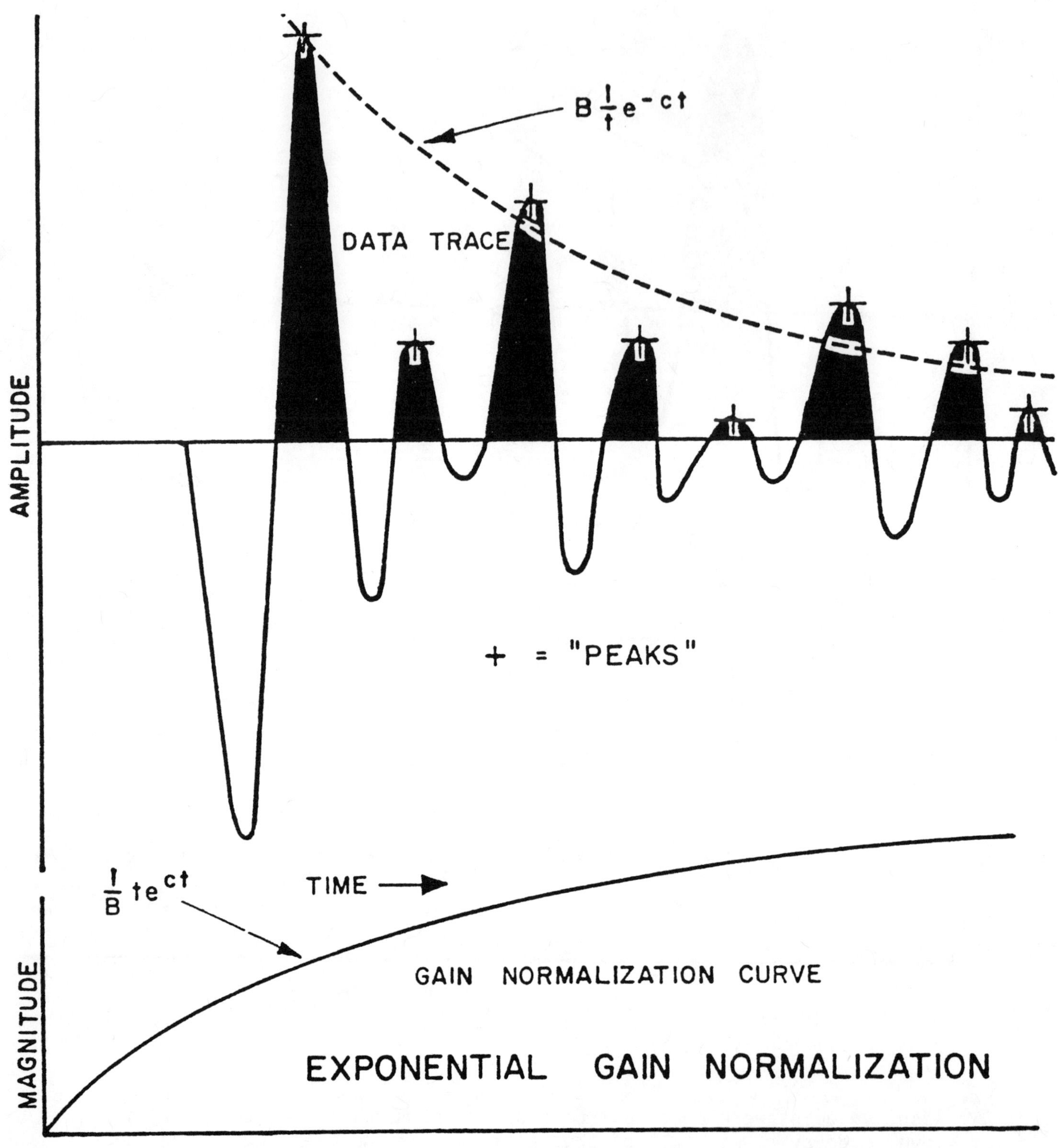

Programmed gains are illustrated by another figure from Lindseth. In this case a gain function is computed universally from all the traces in a profile or even over an entire survey area, and then a reciprocal gain function is applied similarly to every individual data trace. In the figure we note the application of a programmed gain function of exponential form. Because programmed gain functions are slowly varying, no event distortion is introduced with their use.

With the application of programmed gains event sizes retain interpretive significance both laterally (which is readily appreciated) and also over recorded time, although some experience is needed to make these judgements. Standard sections cannot be used in this same way. They tend to amplify all events to a more uniform level, thus ruling out lateral discrimination and most certainly any discrimination in the time sense.

Before passing on, we should not lose our perspective. Adaptive gains or digital AGC's are still of great value and even preferable for producing processed sections in which the entire subsurface geometry is to be defined and interpreted. In fact, it is customary in most approaches to amplitude processing to offer an AGC section as comparison to the amplitude processed section.

C. Controlled Amplitude and True Amplitude Processing

Controlled Amplitude sections seek only to preserve the relative significance of signal level laterally and also in some sense over recorded time. These sections are typically produced using programmed gains as described. True amplitude sections seek to accomplish still more by preserving also an estimate of the true amplitude including all loss effects. Such information requires an innovative form of display owing to the extreme range of amplitude values which must be encompassed, as well as more sophisticated computations.

True amplitudes can be determined by calibration methods which may, for example, make use of a recorded direct arrival from a source of known distance and energy output. They may also be determined from the seismic data using decomposition techniques. Decomposition methods seek to separate all the loss mechanisms and isolate also the reflective strength of each event.

Displays of true amplitude information include use of color superimposed on an AGC version of the processed section and plots of the signal level referred to interpreted reflectors within the section as shown in the following figure from Petty (now Petty-Ray Geophysical Company).

The added information content of true amplitude sections as of this time has not proved significantly greater than that provided by relative amplitude sections. In consequence, the added cost of such techniques frequently limits their use to situations of special study.

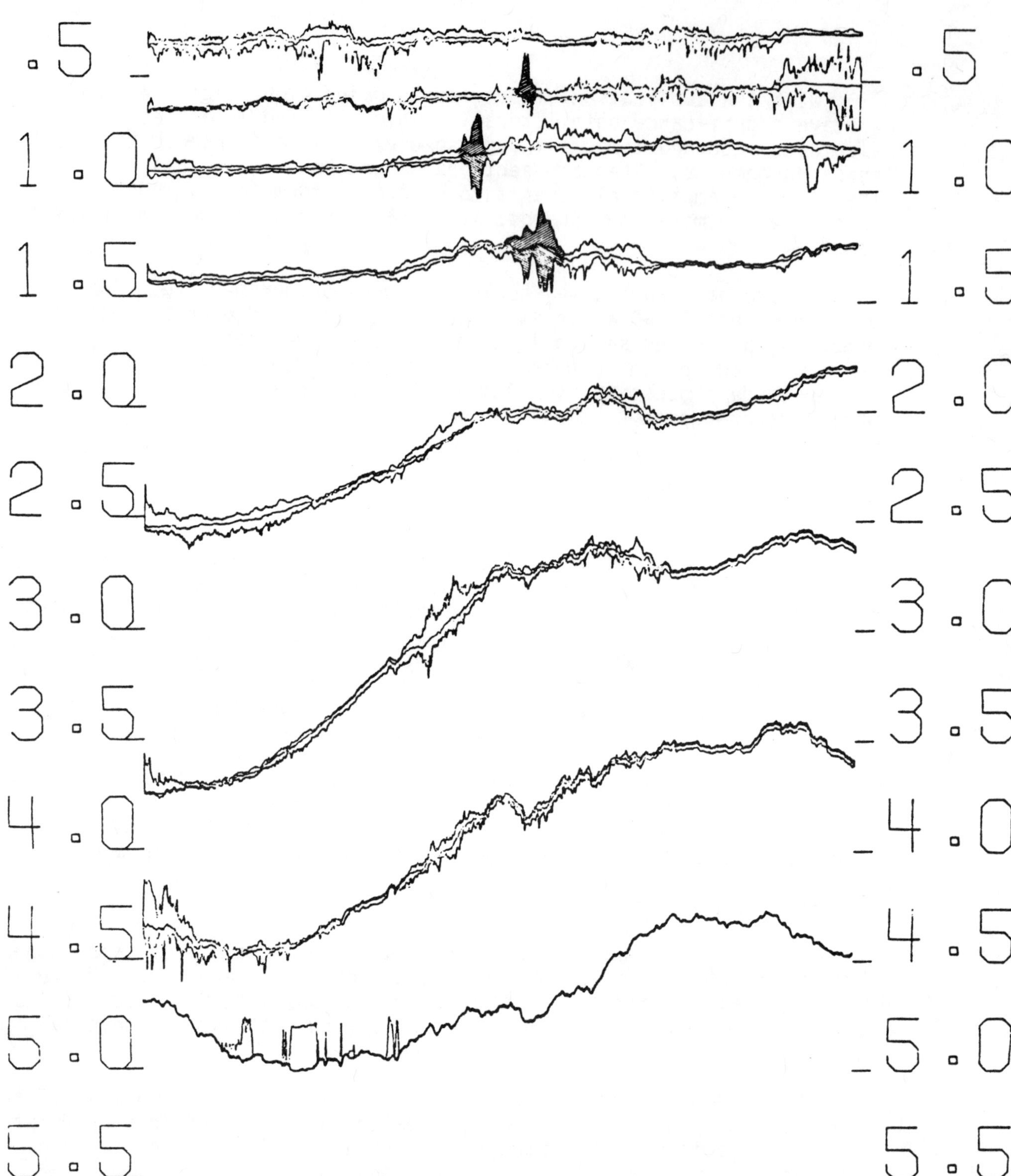

(ESP) EXTRACTION OF SEISMIC PARAMETER

Display of RMS Signal Level for Interpreted Intervals, Horizontally Compressed Display Quantitative Reference Deleted.

(Courtesy of Petty Geophysical Engineering Company)

D. Examples of Alternate Approaches

The following figures are examples of amplitude processing by Teledyne and GeoCom. In both cases we see quality examples of amplitude anomalies which have interpretive significance, and particularly of stratigraphic nature.

Utilization of amplitude processed sections and related direct detection technology has accounted for many discovery wells in the Gulf of Mexico, the Niger Delta, the pinnacle reefs of Michigan, the thick pay zones of the North Sea as well as many other areas both on land and in the shallow marine environment throughout the world. Of course there are many dry holes as well to attest to misapplictions of new technology. On balance, however, the usefulness of amplitude processing is now unquestioned and well established.

Our final illustration notes an instance of amplitude processing of high frequency digital data. Several remarkable features are exhibited including a gas seep along a fault venting at the surface, as well as shallow gas-filled sands. In fact, such results make it clear that there are still further uses of seismic amplitudes which remain yet to be exploited.

PROCESSING COMPARISON
HIGH FIDELITY vs. STANDARD
(Courtesy Of Teledyne Exploration Company)

HIGH FIDELITY PROCESSING

STANDARD PROCESSING

NOTES

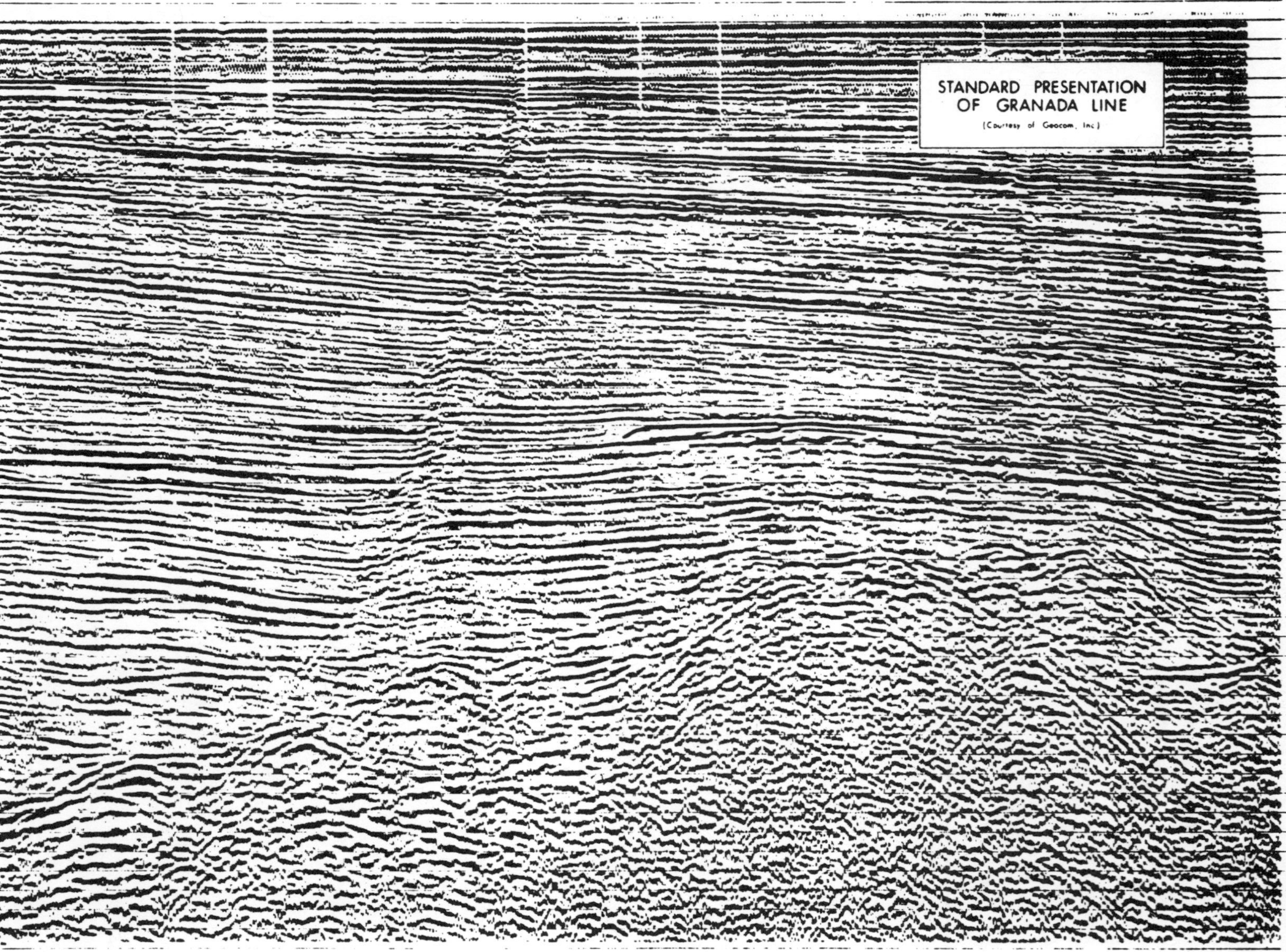
STANDARD PRESENTATION
OF GRANADA LINE
(Courtesy of Geocom, Inc.)

SOUTHEAST

CONTROLLED GAIN SECTION OF GRENADA LINE

(Courtesy of Geocom, Inc.)

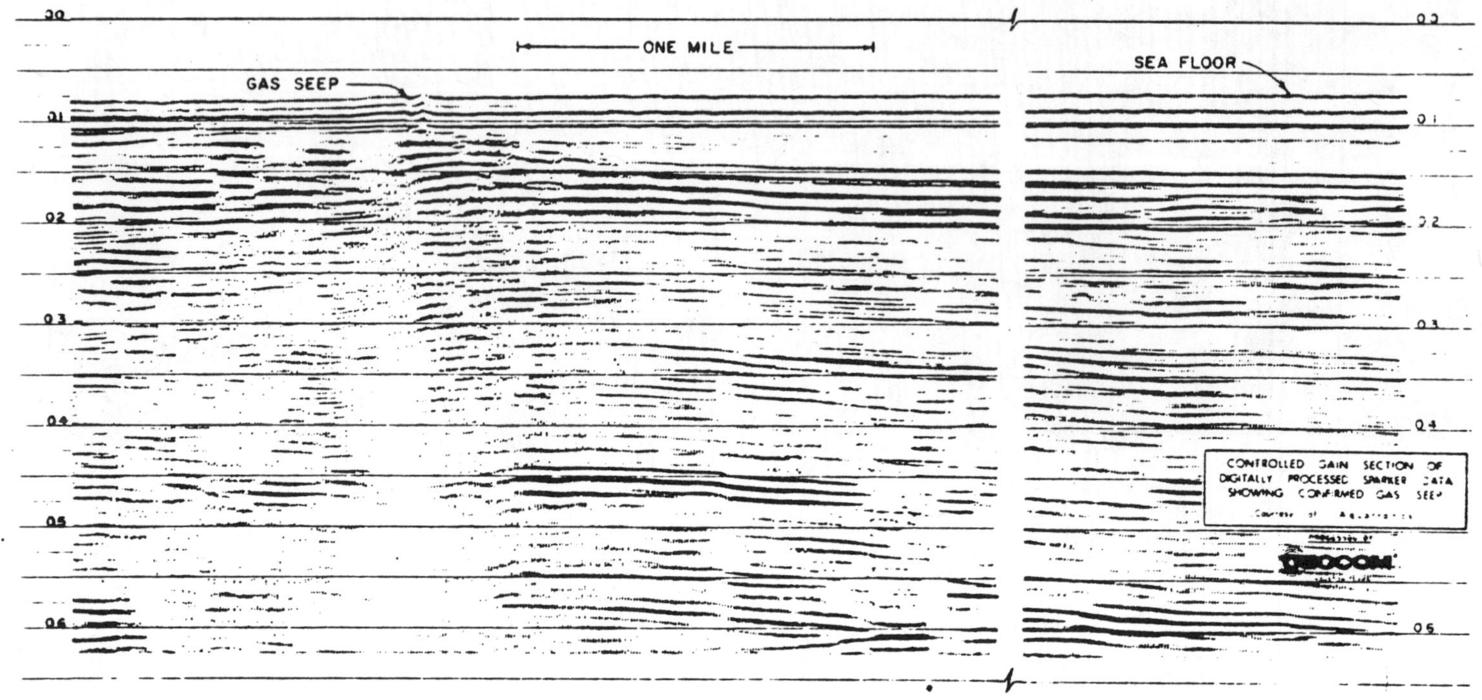

IV. The Role of Seismic Wavelets and Wavelet Processing

A. Waveforms, Frequency Content and Interpretive Significance

Use of waveforms is a natural extension of using seismic amplitudes and also a part of our approach to interpretation in the ideal case. Waveforms, in fact, include both amplitude information as a function of frequency and phase as function of frequency. Phase is a rather subtle concept which relates to the alignment in time of the frequency components.

We first consider an exhibit provided by Lindseth and Petty-Ray Geophysical Co. which shows the amplitude content of a complete seismic record and a single Ricker wavelet as functions of frequency. Much of our thinking has been directed toward the precept that the resolution potential, and hence the interpretive potential of a seismic waveform rests with the broadness of its frequency band. Hence, waveforms which encompass much high frequency must permit highly resolved interpretations. While these ideas are true, they do not express the entire circumstance. The shape of the waveform and its corresponding phase strucure are of equal if not of greater interpretive significance.

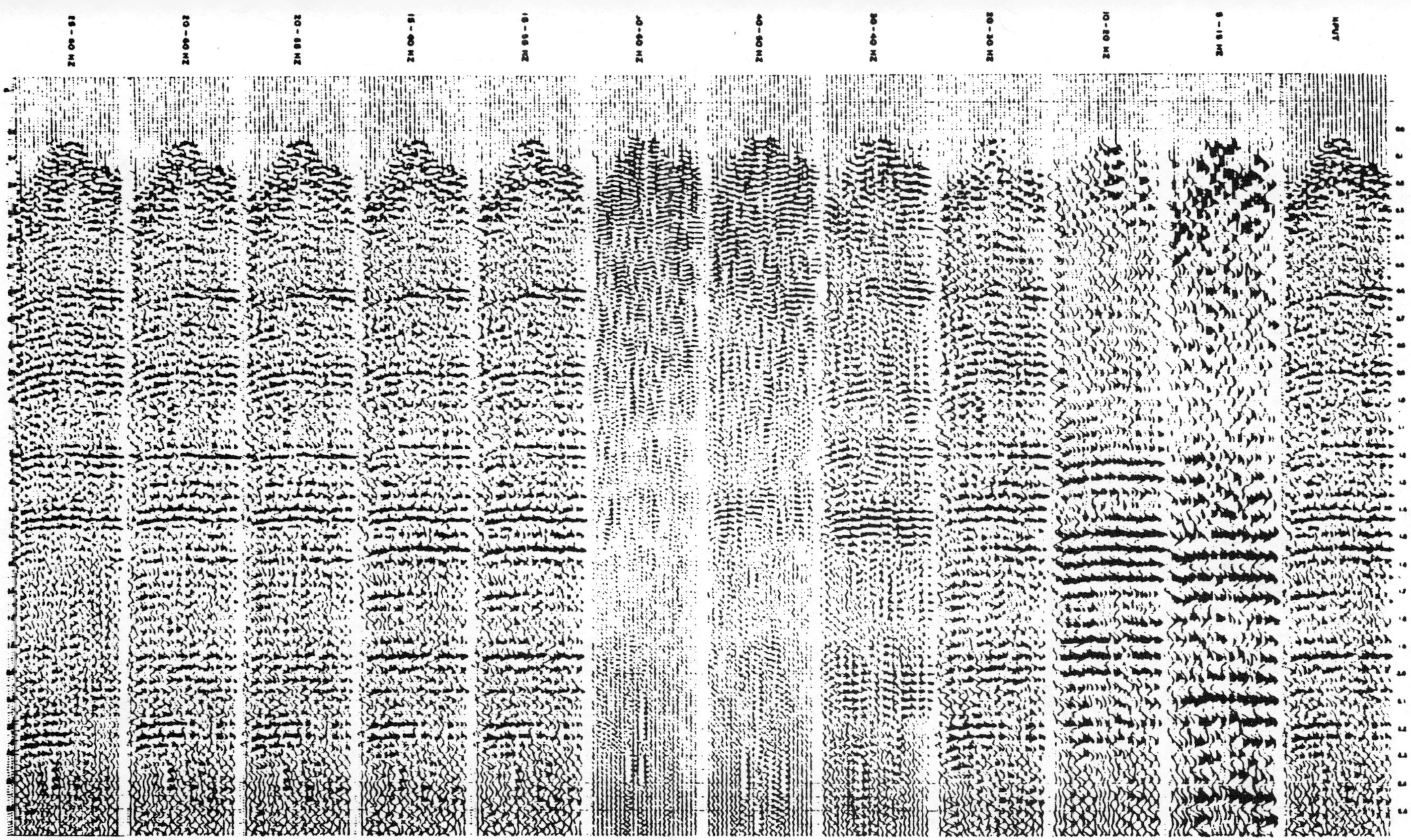

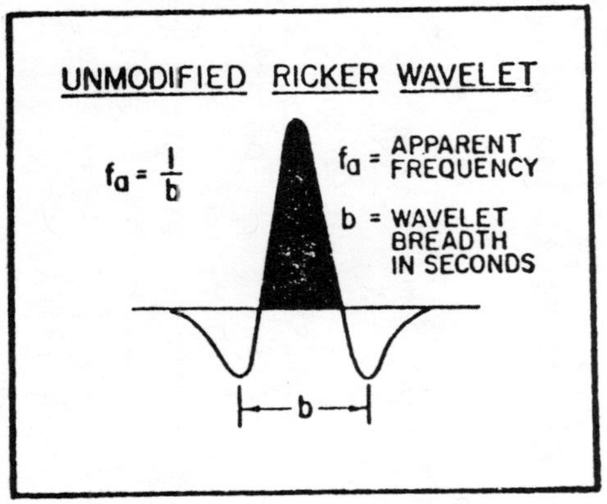

UNMODIFIED RICKER WAVELET

$f_a = \frac{1}{b}$

f_a = APPARENT FREQUENCY

b = WAVELET BREADTH IN SECONDS

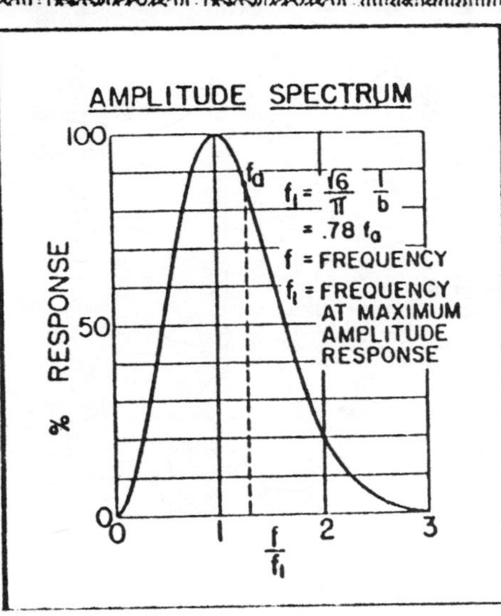

Decomposition of a seismic record and a seismic trace into frequency components.

We first encountered the significance of waveform shape in developing the interpretive scheme for an ideal subsurface. Where the zero phase-symmetric waveform was present we could identify individual reflection events and accomplish much that was not possible with a more complex waveform. Schoenberger of Exxon (1974) and Berkhout (1973, 1974), who was with Shell at the time, both recognized the merits of the simple waveform and were even able to express this in theoretical terms.

In the figure which follows we outline the transformation of a seismic waveform to zero-phase symmetric form. We note here a case in which the Basic Marine Waveform and the symmetric resultant waveforms differ mainly in their phase properties. The interpretive advantage inherent in the symmetric waveform is quite apparent.

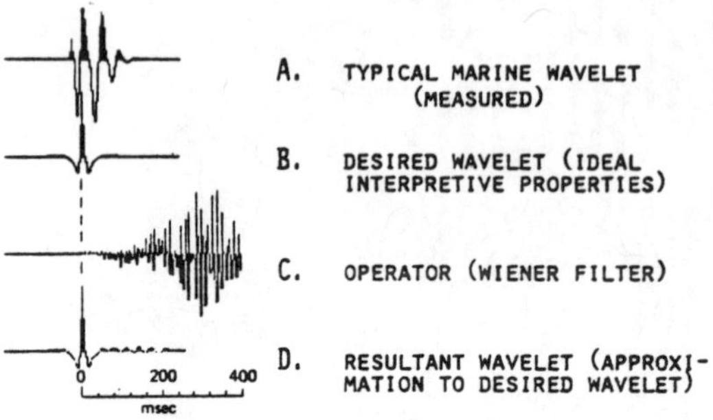

—Wavelet signature-processing transformation of basic marine wavelet to ideal interpretable form by Wiener filtering.

Wavelet processing as portrayed by the figure is quite distinct from the conditioning of waveforms via deconvolution in several key regards. Although both deconvolution and wavelet processing employ the mathematics of Wiener filters, wavelet processing seeks to work in the frequency band present in the data and makes no statistically based assumptions. Deconvolutions seek to broaden the effective bandwidth and take as assumptions a random character for the underlying reflectivity series in the data and a minimum phase nature for the waveform. In the face of such assumptions deconvolutions are far more likely to produce poor results than wavelet processing. On the other hand, wavelet processing requires an estimate of the propagating waveform and we must discuss how these are to be obtained.

B. Estimation of Seismic Waveforms and Wavelet Processing

Since wavelet processing as just described always requires an estimate of the propagating seismic wavelet, some discussion of means for obtaining estimates is needed. First, however, we should justify our unstated assumption that the propagating wavelet in a stacked seismic trace may be said to be largely invariant over all of the recorded time.

If we examine carefully the physics of the reflection and transmission process, it becomes clear that except for attenuation and dispersion, no other mechanism changes the waveform other than by a simple scaling of its amplitude. Apparent changes in form do result in the seismic trace where we regard superimposed waveforms which can result from closely spaced reflecting boundaries, interbed multiples or the filtering effect of transitional boundaries, but these are in fact only apparent.

Turning to attenuation and dispersion and investigating them more closely further suggests that on the normal plotting scales for seismic sections and for usual interpretive procedures, these waveform distortions may be overlooked and regarded as second-order effects. No doubt the refined methods to be developed in the future will take advantage of these differences as yet another factor by which to identify lithology.

With an approximately constant waveform in the individual seismic trace we can now consider means and methods for determining its character. Three principal techniques categorize the approaches which are currently in use. First we may attempt direct measurement. Next we may adapt a deterministic approach and finally we may attempt statistical estimation.

(1) Direct Measurement of Seismic Waveforms

Measurement of uncontaminated seismic waveforms is only feasible in the deep marine environment. Here a good approximation to a homogeneous half-space truly exists. The measurements may be made up close to the source where they are termed "near field", or at some distance comparable to those of reflected events, thus constituting "far-field results". In all instances of measurement, the effect of surface ghost reflections, array induced filtering effects, instrumentation filters and cable effects must all be accomodated to produce a valid representation of the waveform which is to be found in reflections.

In the figure which follows, a measured near-field waveform labeled as "C" is shown. Corrections have been applied for the surface reflection or ghosting effects so that the waveform should match the far field one which propagates in the subsurface. Wavelet processing with the waveform"C"gives the zero-phase symmetric waveform shown as "E".

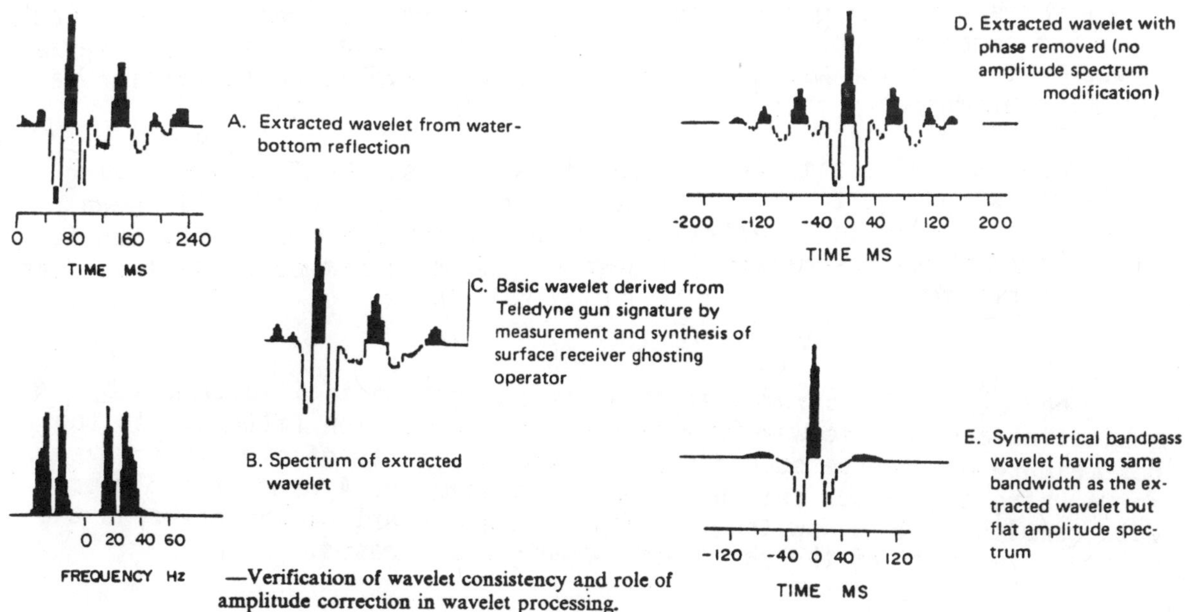

—Verification of wavelet consistency and role of amplitude correction in wavelet processing.

(2) Deterministic Wavelet Extraction

If we start with a multiple-free, relatively noise-free seismic trace where the ideal single trace model may be applied, we can use a deterministic approach to wavelet extraction. In this case the seismic trace may be regarded as a convolution or filtering of the reflectivity series by the waveform. If we determine by independent means the reflectivity series, then it may be removed to reveal an estimate of the waveform.

Examples of independent reflectivity series determinations may consist of computations from well log measurements or reasonable geological inferences. One common case of inference would be to presume that a water-bottom reflection event which was free of refraction arrivals would constitute a good approximation to a single reflection event. This approach is illustrated by waveform "A" of the Figure which has been derived under such assumption. Note how well it agrees with "C", the measured result. It too would produce "E" after transformation. Here then is a verification of the validity of the waveform estimate.

Another example of a deterministic approach might be to infer that a Gulf Coast bright spot consists of a single gas sand having a consecutive minus-plus reflectivity series. Of course the thickness of the sand, or equivalently the time separation of the reflective spikes, would have to be established by some type of analysis procedure.

(3) Statistical Estimation of Wavelets

When all else fails in estimating waveforms we attempt statistical estimation with the assumptions these engender. Of course deconvolution methods fall into this category. For example, if we attempt to design an operator which transforms a seismic waveform into a spike, then we really are attempting to design an inverse wavelet. Hence if we treat a spiking deconvolution operator as an inverse wavelet, then its inverse must in fact be the wavelet. Dr. Robinson's wavelet noted earlier was in fact produced in this way.

Statistical estimation procedures work best when the underlying wavelet is reduced to its simplest form by accounting for all known distortions prior to undertaking wavelet estimation. Compensation for instrument phase distortions is perhaps the most common correction step of such nature.

The nature of instrumentation phase distortion is illustrated by subjecting a symmetric waveform to the instrumentation filters. In the next Figure we observe such distortion for a variety of symmetric waveforms using the filter setting normally applied for 4 msec. and 2 msec. data sampling. Note how the severity of the distortion increases as the central frequency of the particular waveform increases.

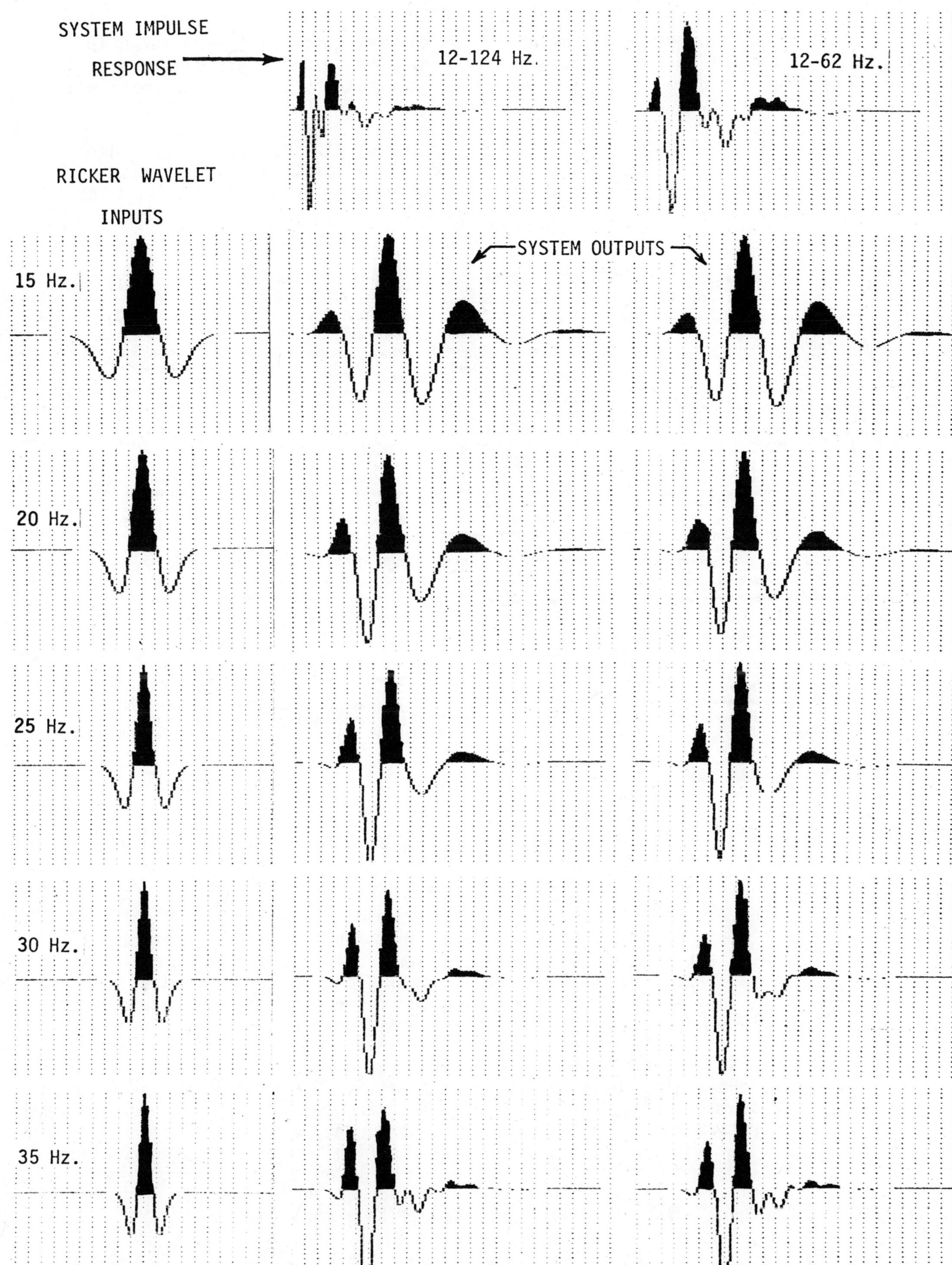

SYSTEM IMPULSE
RESPONSE
12-124 Hz.
12-62 Hz.
RICKER WAVELET
INPUTS
SYSTEM OUTPUTS
15 Hz.
20 Hz.
25 Hz.
30 Hz.
35 Hz.

Dephased panels of data from the Williston Basin are shown in contrast to undephased data. The distinctions in this case are subtle, but a subsequent step of wavelet estimation by statistical means would be improved.

Usual assumptions inherent in the statistical procedures involve the phase structure of the wavelet, the statistical distribution of the subsurface reflectors and similar considerations. These may or may not be valid.

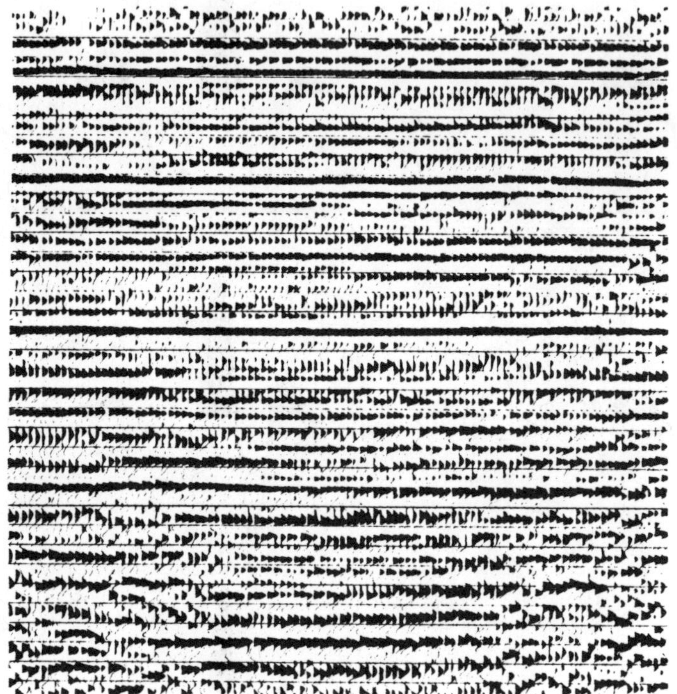

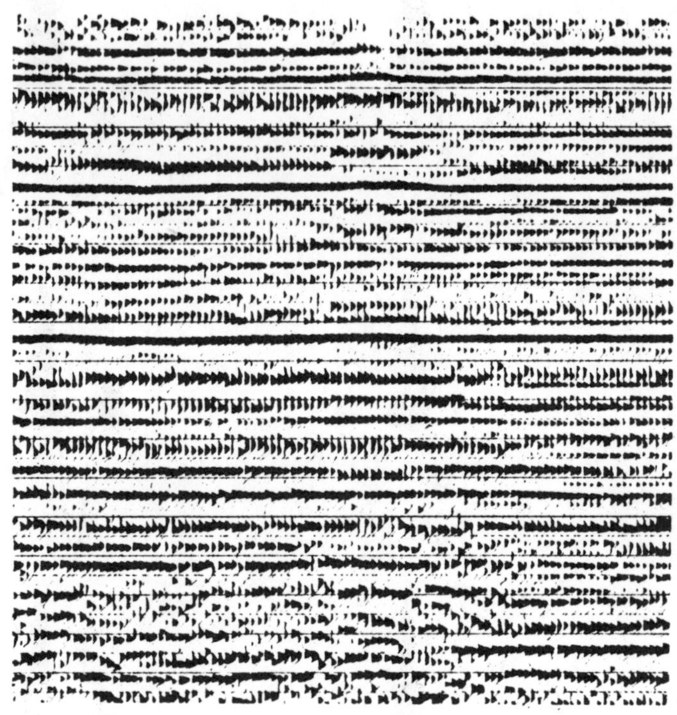

DEPHASED

(Courtesy of Houston Processors, Inc.)

Waveform "D" of the figure prior to the Williston Basin data shows the effect of taking the estimated waveform and correcting only its phase. Note that while there is resolution improvement, we still have a less than ideal waveform. Hence, the correction for the waveform shape must compensate for both phase and amplitude structure.

Several estimated marine seismic wavelets and their processed forms are shown along with the operator spectra. One caution is in order. Waveforms having deep "holes" in their frequency spectra produce shaping operators which greatly magnify the frequency content of such holes. If the reflection sequence of the data is rich in such frequency components, application of the shaping operators will cause a severe "ringing" condition. Shaping operators having such a characteristic must be "detuned" prior to their use to avoid this contingency.

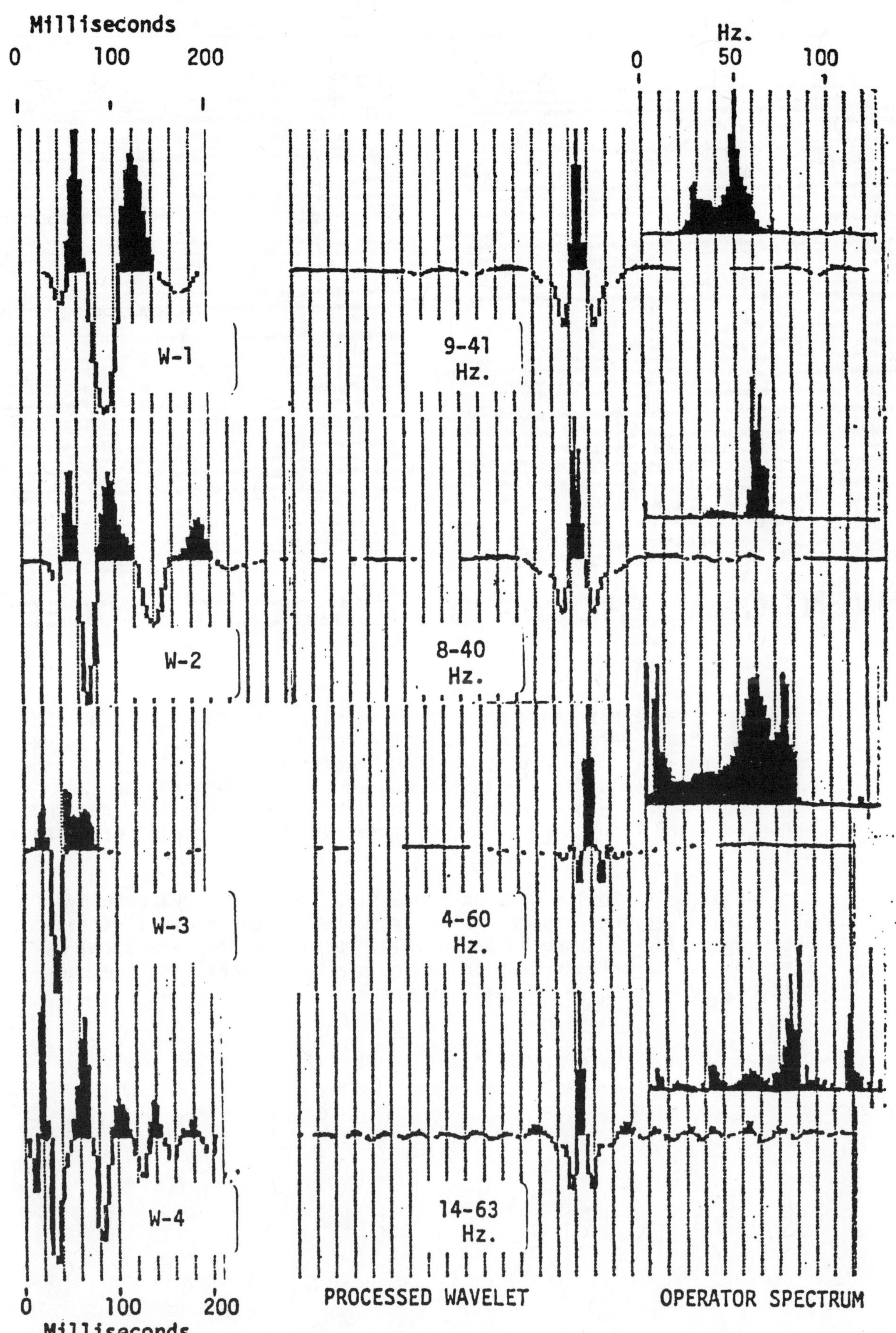

Milliseconds
0
100
200
Hz.
0
50
100
W-1
9-41
Hz.
W-2
8-40
Hz.
W-3
4-60
Hz.
W-4
14-63
Hz.
0
100
200
Milliseconds
PROCESSED WAVELET
OPERATOR SPECTRUM

CONVENTIONAL PROCESSING

WAVELET PROCESSING

The data panels above show the wavelet processed version of the Williston Basin data in contrast to the conventional data. These examples were again provided by Houston Processors, Inc. Since we know little of the subsurface in this area, we can comment only on the improved continuity of events, the consistency in character, and the enhanced resolution which is clearly present in the wavelet processed data.

An example showing wavelet processing in central Oklahoma by Seismograph Service Corporation is presented in the next illustration. Below each stacked trace is the wavelet recovered by statistical estimation for that trace. While there is some variability, the consistency we do observe is quite remarkable.

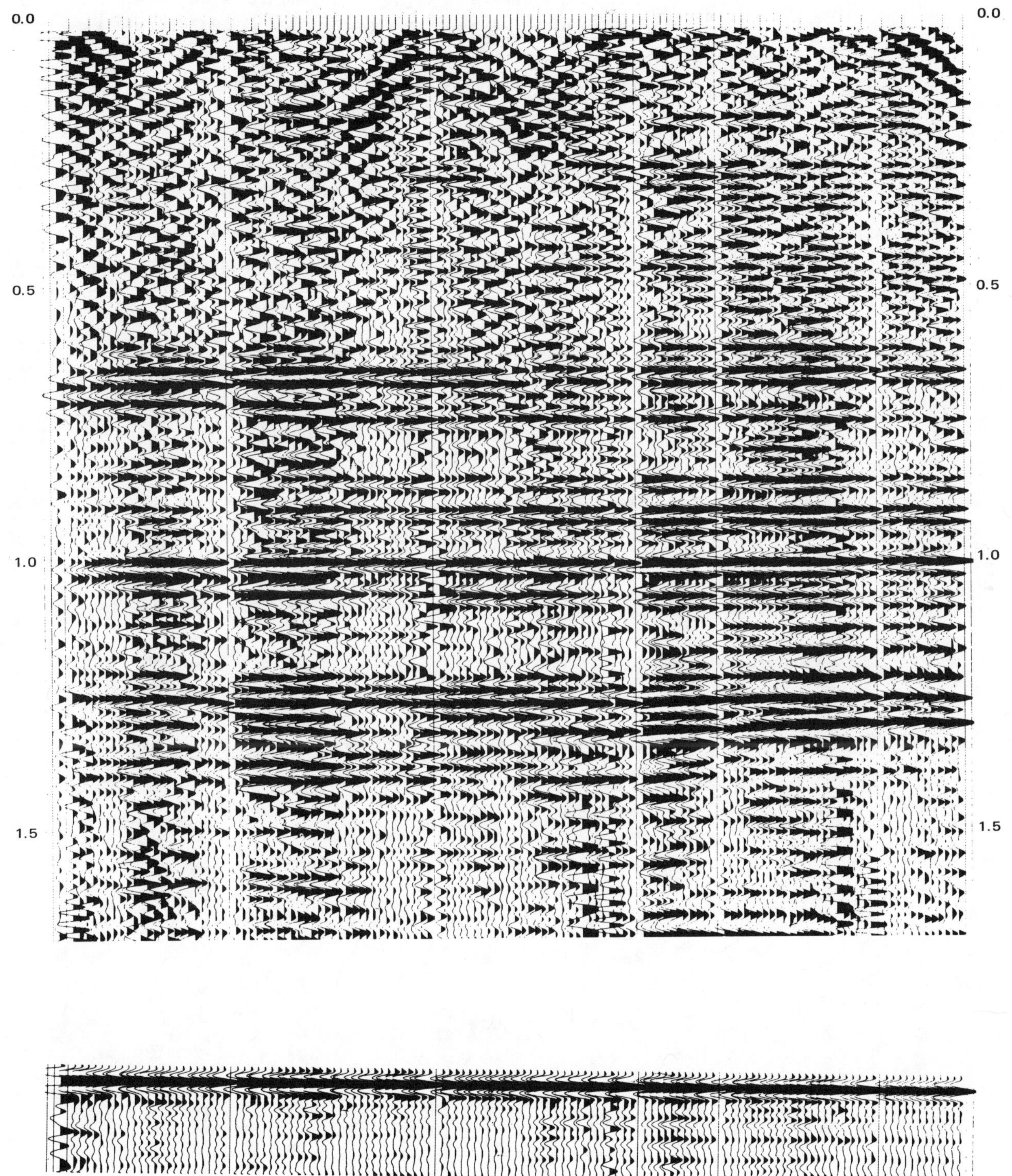

(Courtesy of Seismograph Service Corporation)

(4) Detailed Case Study

The following sequence of eight figures document a marine wavelet processing case study performed by GeoQuest International, Inc. In this treatment the instrument dephasing step is accomplished separately. Note how dramatically it simplifies the data.

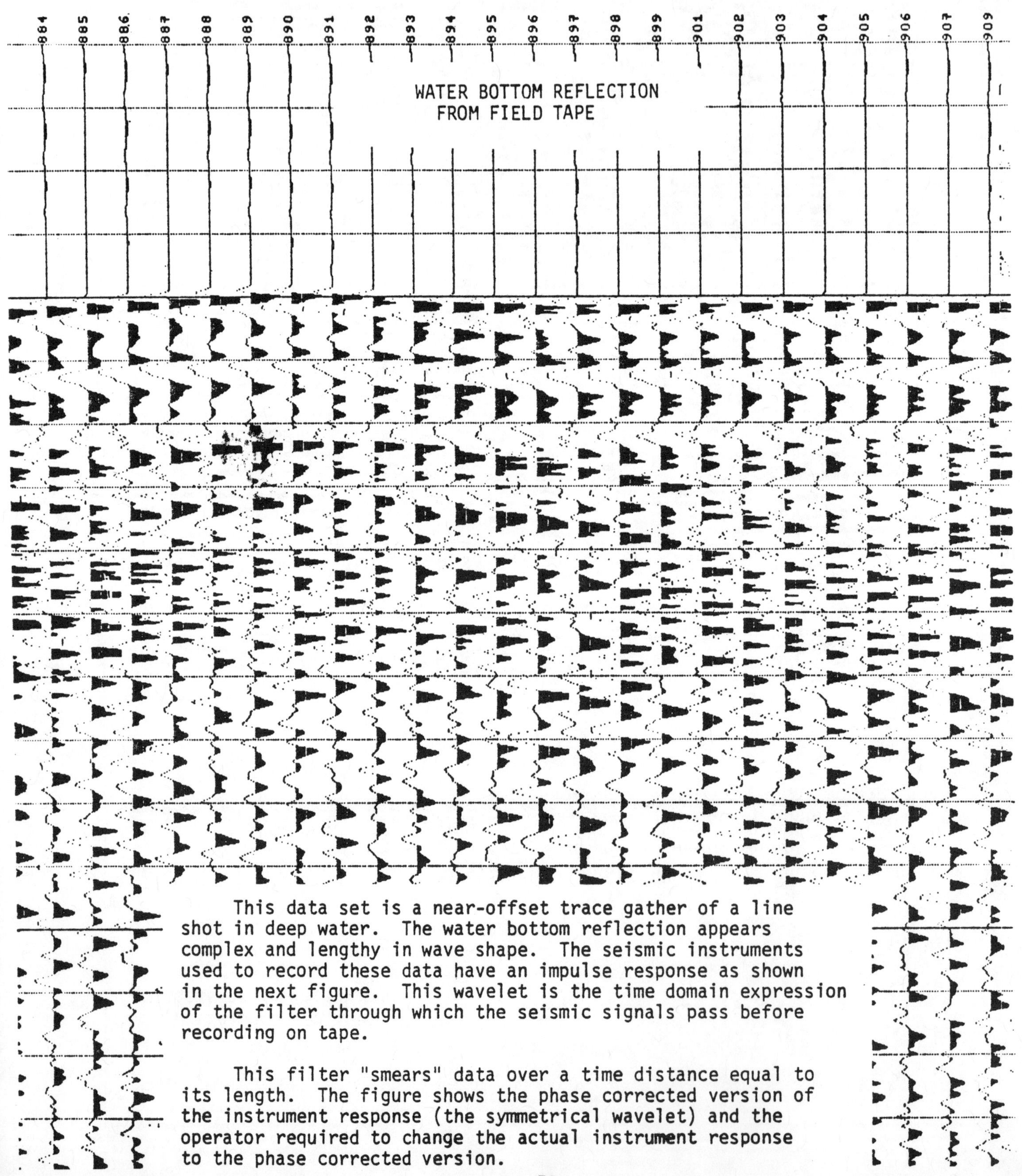

This data set is a near-offset trace gather of a line shot in deep water. The water bottom reflection appears complex and lengthy in wave shape. The seismic instruments used to record these data have an impulse response as shown in the next figure. This wavelet is the time domain expression of the filter through which the seismic signals pass before recording on tape.

This filter "smears" data over a time distance equal to its length. The figure shows the phase corrected version of the instrument response (the symmetrical wavelet) and the operator required to change the actual instrument response to the phase corrected version.

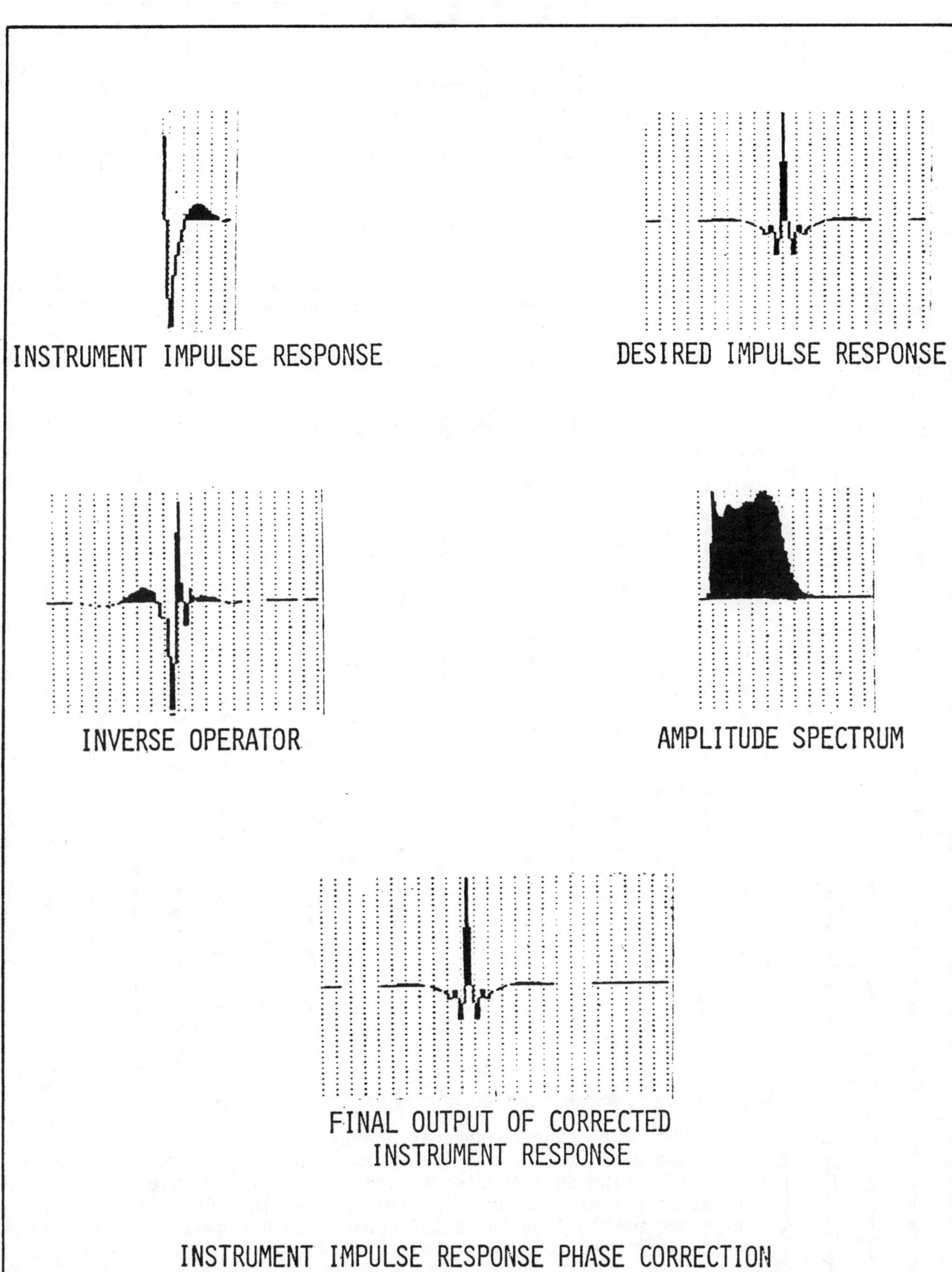

INSTRUMENT IMPULSE RESPONSE PHASE CORRECTION

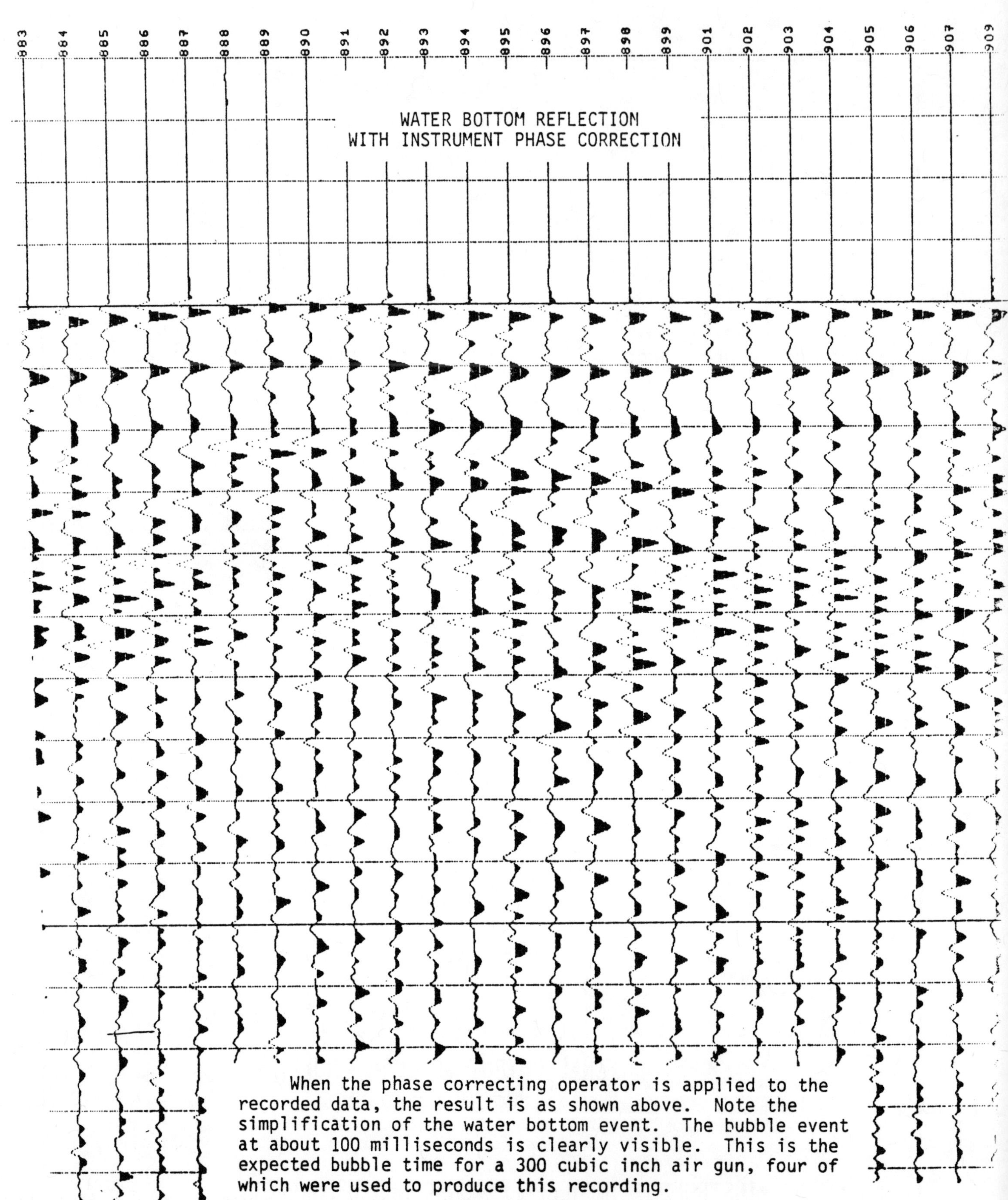

When the phase correcting operator is applied to the recorded data, the result is as shown above. Note the simplification of the water bottom event. The bubble event at about 100 milliseconds is clearly visible. This is the expected bubble time for a 300 cubic inch air gun, four of which were used to produce this recording.

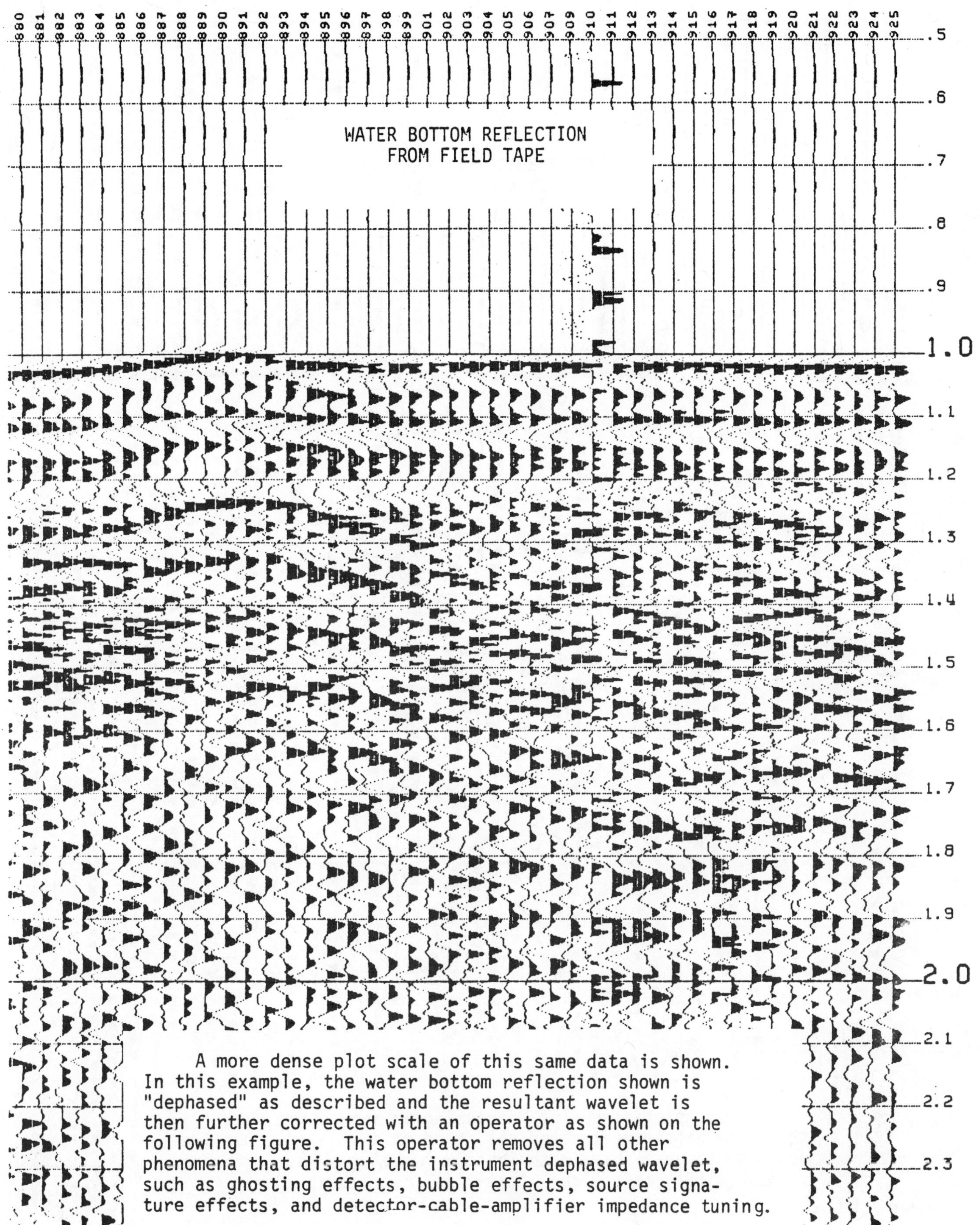

A more dense plot scale of this same data is shown. In this example, the water bottom reflection shown is "dephased" as described and the resultant wavelet is then further corrected with an operator as shown on the following figure. This operator removes all other phenomena that distort the instrument dephased wavelet, such as ghosting effects, bubble effects, source signature effects, and detector-cable-amplifier impedance tuning.

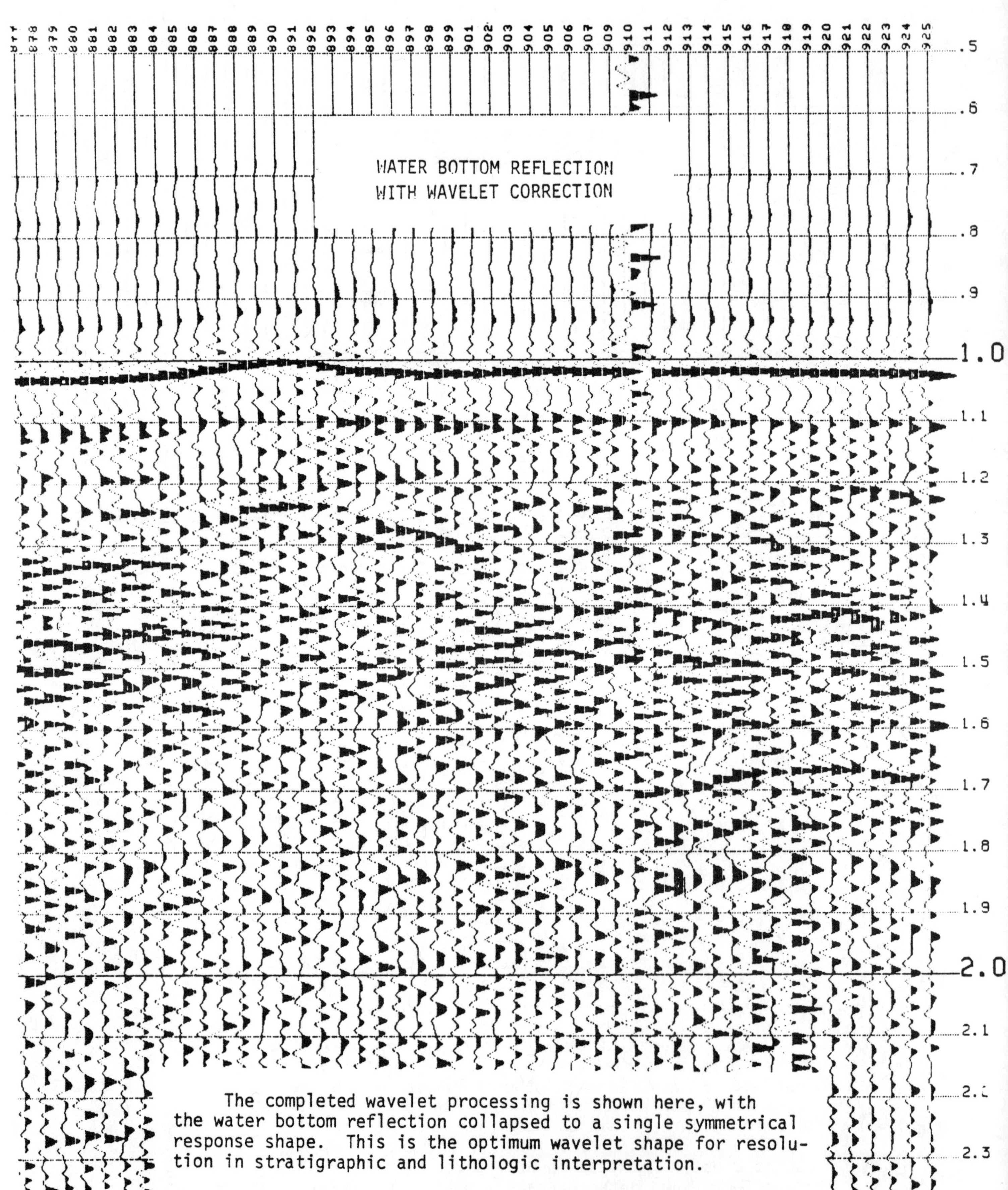

The completed wavelet processing is shown here, with the water bottom reflection collapsed to a single symmetrical response shape. This is the optimum wavelet shape for resolution in stratigraphic and lithologic interpretation.

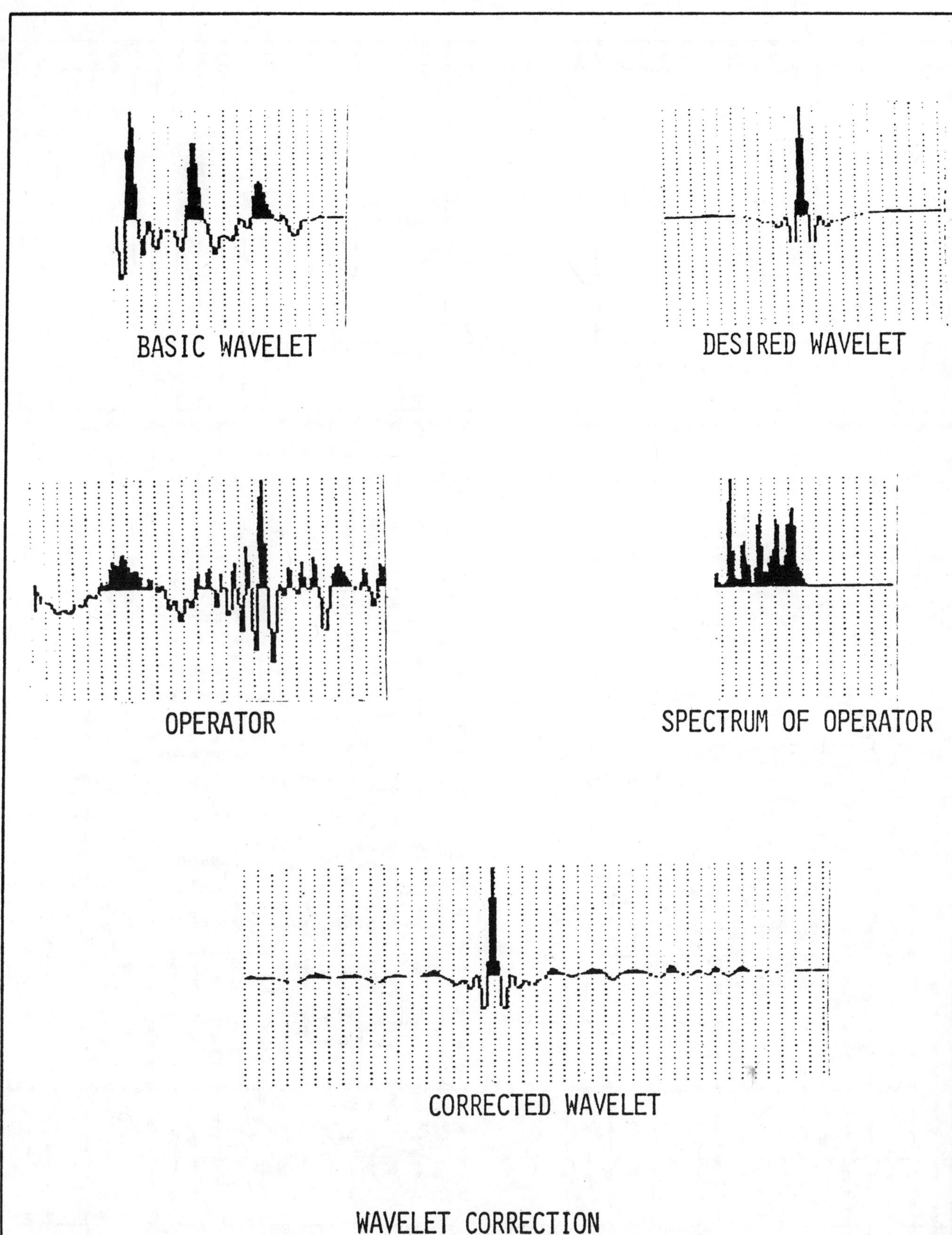

WAVELET CORRECTION

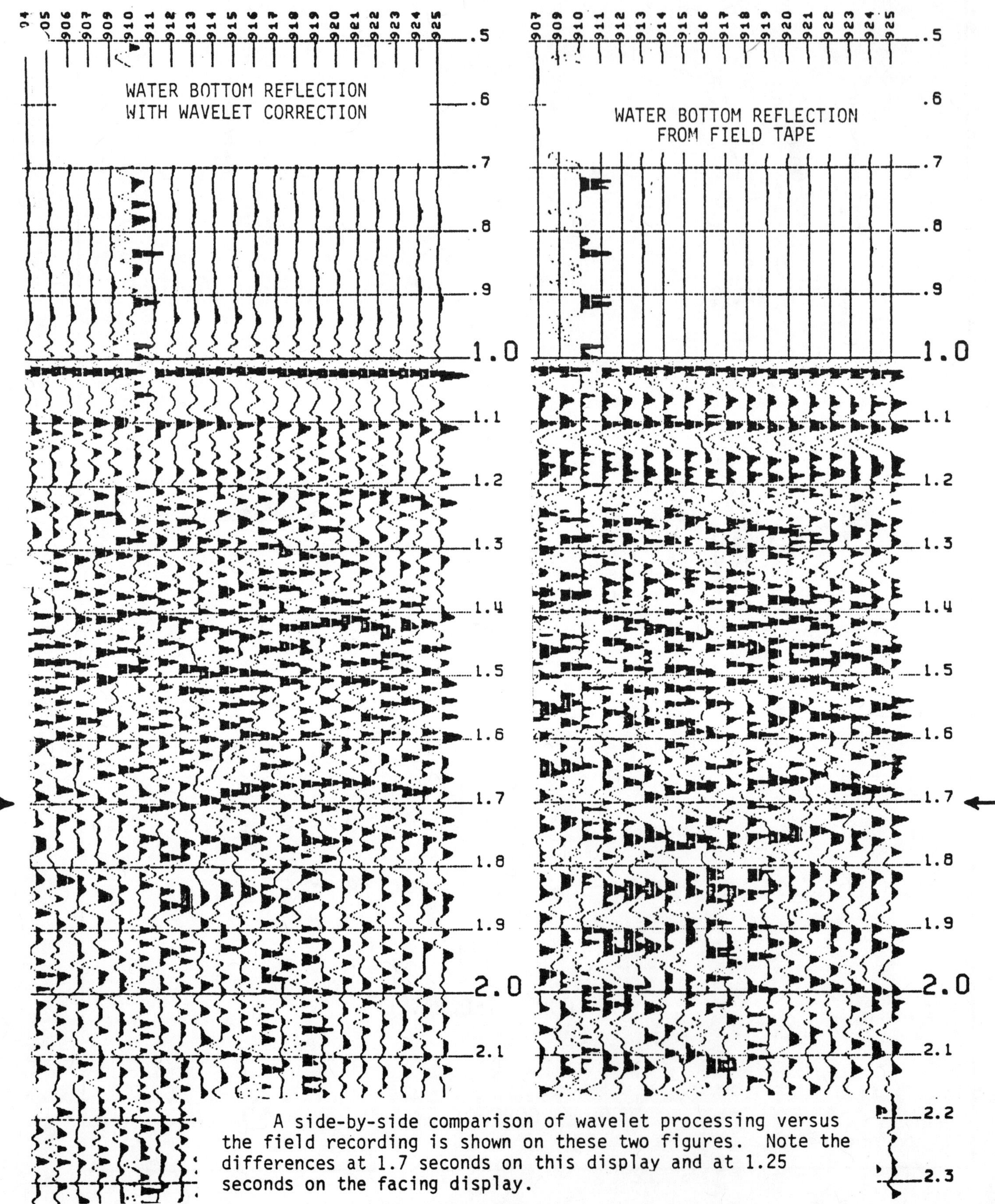

A side-by-side comparison of wavelet processing versus the field recording is shown on these two figures. Note the differences at 1.7 seconds on this display and at 1.25 seconds on the facing display.

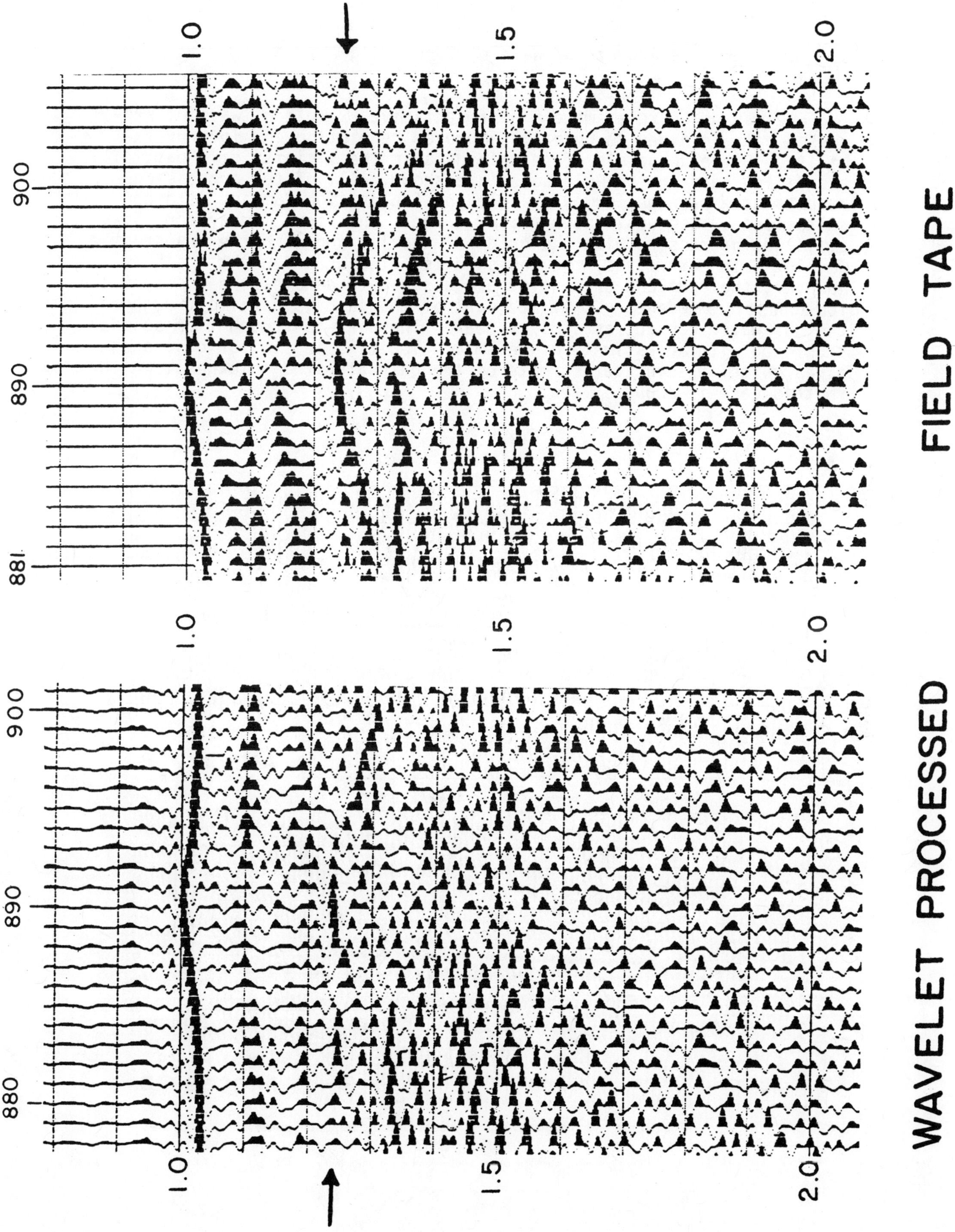

FIELD TAPE

881 890 900

1.0 1.5 2.0

WAVELET PROCESSED

880 890 900

1.0 1.5 2.0

C. Interpretive Aspects of Wavelet Processing

We may go on to examine a representative example of wavelet processed data in order to note its improved resolution and interpretability. Close-up panels of seismic data in the vicinity of a partially gas-filled sand in the Gulf Coast are shown. All of the processing results have been performed by Teledyne Exploration Company with consultations by GeoQuest International, Inc. The first comparison shows amplitude processing with the original wavelet and with the transformed wavelet. In both cases, the gas-sand is noted as a "bright spot", however, the "SHAPER" or wavelet processed data denotes the onset of the low acoustic impedance gas-filled part of the sand as a white trough and the gas-water contact as a black peak. The peak-to-trough time separation used with an estimate of seismic velocities in similar gas sands gives quite an accurate estimate of the gas-saturated thickness, which in this case is about 65 feet.

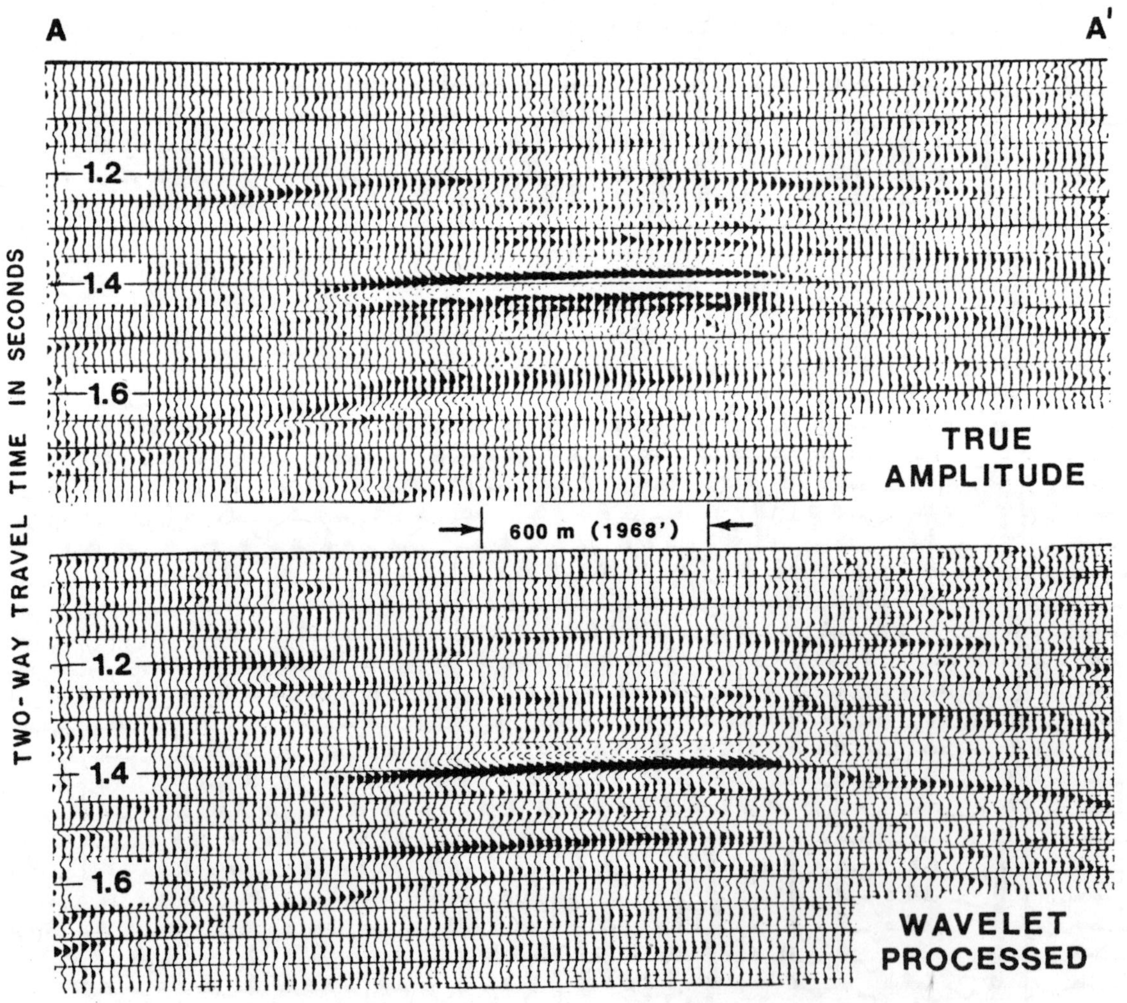

—Comparison of conventional and wavelet processing, amplitude preserved.

The differences between the conventional waveform section and the wavelet processed section go far beyond a simple reversal of display polarity as might be suggested by a first glance. In the wavelet processed data, waveshape as well as amplitude is a most significant interpretive diagnostic. Hence, we may next look at the same panels of data with conventional display scalings and now note other low acoustic impedance zones in the wavelet processed result. These correspond to water-filled sands and do not have the high amplitude characteristic of the gas-filled sand.

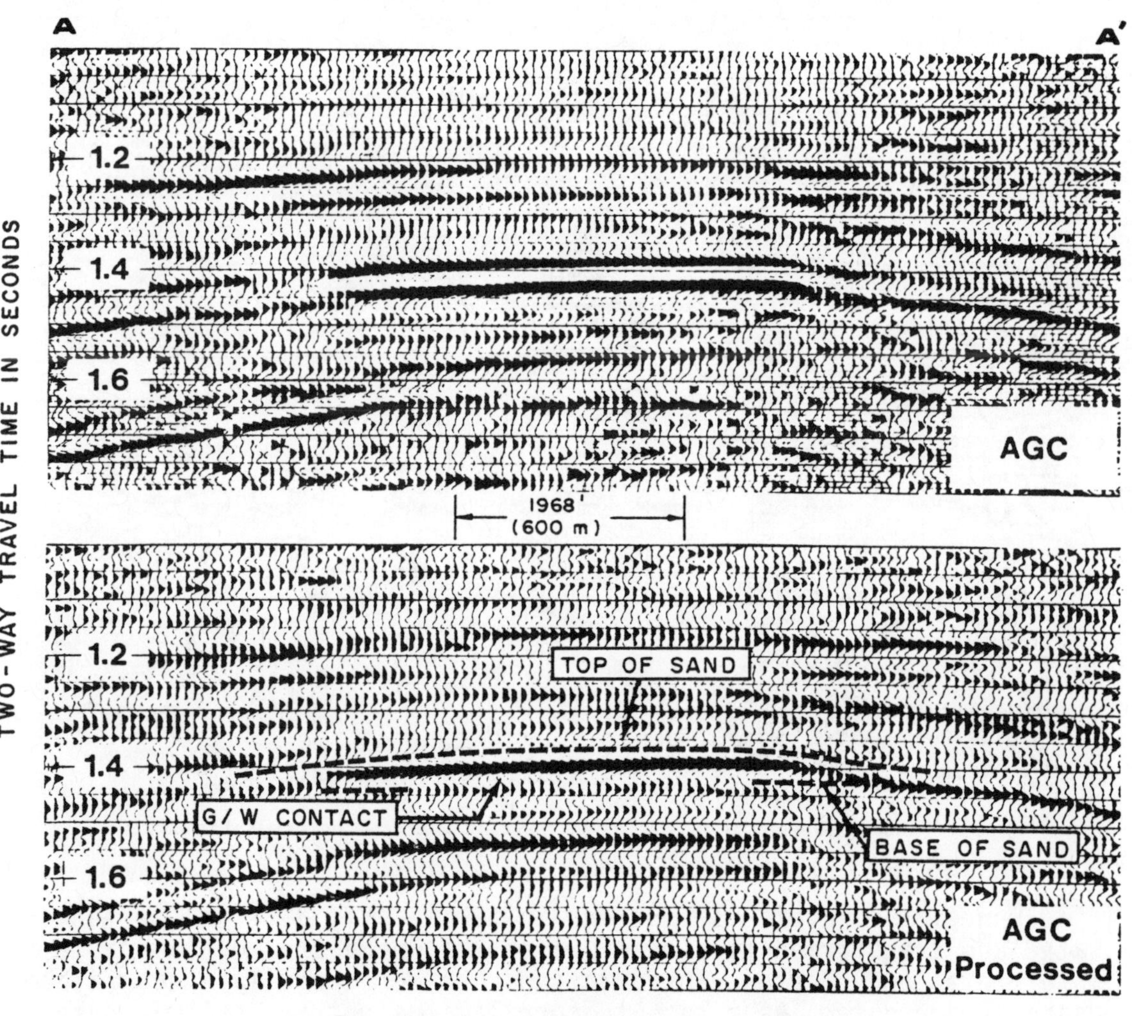

—Comparison of conventional and wavelet processing, AGC.

Dedman, Lindsey and Schramm (1975) interpreted most of the sands in the subsurface column between 1.0 and 1.8 seconds from the seismic data using the principles described. Similarly, sands were identified by means of well log measurements and these were plotted on a two-way travel time scale for direct comparison with the seismic data. The results are shown in the final figure of this series. In assessing the remarkable agreement, it is important to note that the top of a sand corresponds in each case to a trough or white event, while the base is defined by a peak or black event.

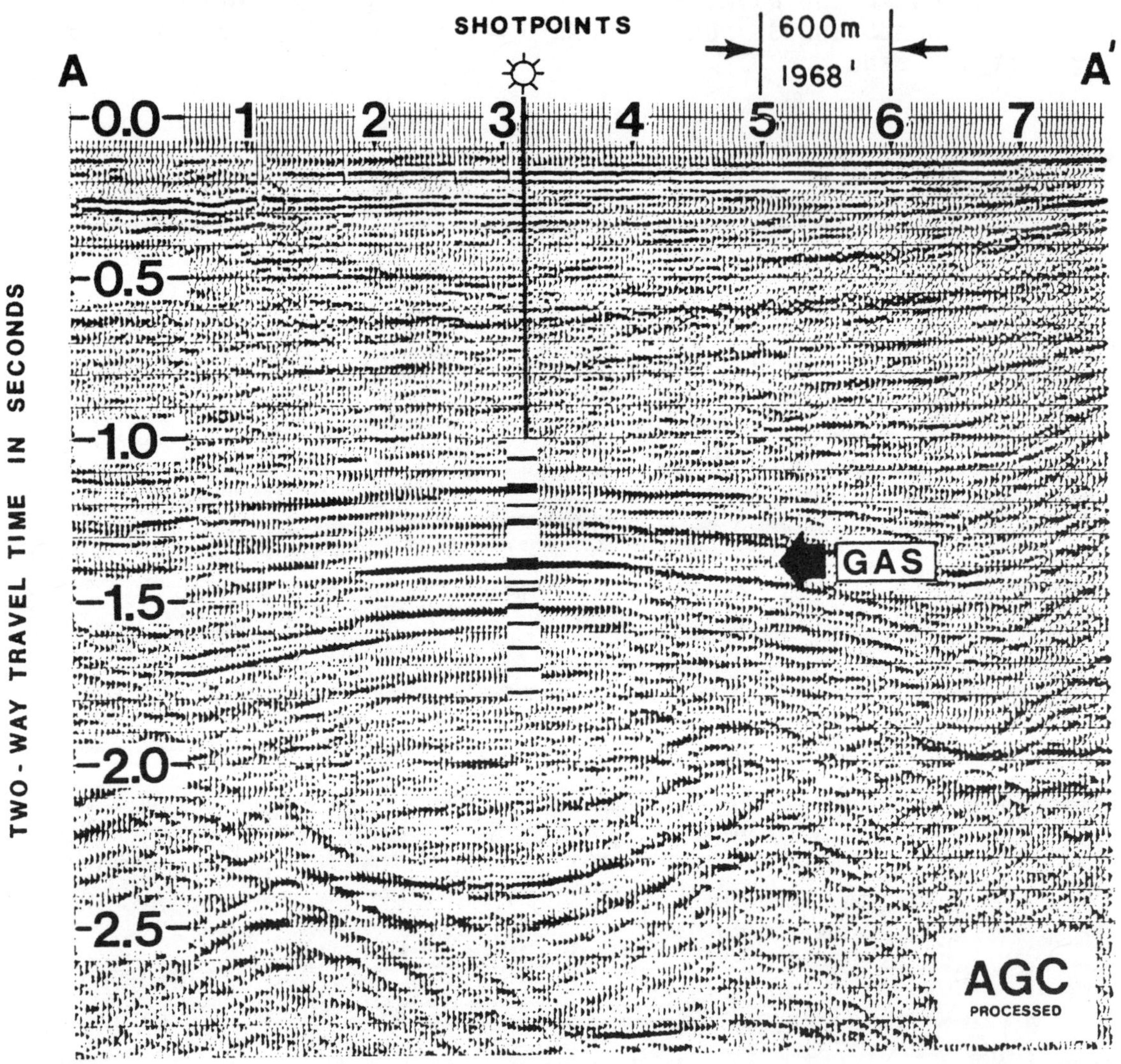

—Lithologic-log overlay on AGC wavelet-processed data. Sandstones marked in black.

Hence, the waveforms inherent in our seismic data can be manipulated and transformed to simpler waveforms which are more amenable to interpretation. These transformations are more readily accomplished using marine data. As a result, correlations between seismic data and geological inputs can be affected with greater rigour to produce more definitive results as indicated here and in previous discussions. We may thus go on to consider the treatment of other problems and matters which were brought up in our initial discussion of the relation of seismic data to stratigraphy.

V. Contributions of GeoSeismic Modeling

A. Introduction

Seismic modeling is in fact the successor to the synthetic seismogram and is also the vehicle by which seismic correlations with geology may be verified. Although publications describing modeling techniques and applications are few in number (see Taner, Cook and Neidell (1970) and Shah (1973), the technology has reached a rather sophisticated level. The paper by Neidell (1975) describes some of the more advanced considerations and limitations associated with the use of seismic modeling.

Unlike the synthetic seismogram, the true seismic model accepts a description of a subsurface in terms of its geometry and acoustic parameters including the usual velocity, density and an attenuation factor. In two-dimensional modeling systems which view the subsurface as having perfect lateral continuity and homogeneity outside of the plane in which the subsurface is described, geometry of virtually any complexity may be treated. Further, the seismic parameters are permitted to vary both horizontally and laterally to represent lithologic transitions and similar subtle stratigraphic effects.

In our introductory discussion we briefly viewed simple modeling illustrations using ray trace approaches and wave theory. These portrayed modeling as a tool for overcoming geometric effects on the seismic section which might obscure stratigraphic objectives and also for treating diffractions, another member of the class of seismic events which do not simply relate to lithologic contrasts. Here we wish to look at modeling from a more advanced viewpoint. In particular we will use it to develop insights into the spatial resolution inherent in seismic data and the very origin of reflections in a three-dimensional subsurface. We shall further note model studies in a more practical context - the generic characterization of specific exploration targets.

B. Sand Identification

We first consider a geologically conceived model taken from Dedman, Lindsey and Schramm (1975). Here the scenario for a barrier bar sand is developed entirely in geologic terms. The purpose of the study is to characterize barrier bar sands as an exploration objective for seismic data as distinct from other exploration targets such as channel sands. A wave theory response is calculated for the model using typical Gulf Coast parameters.

Two waveforms are applied to produce seismic sections. The typical marine waveform gives a section which is characterized by interfering waveforms suggesting a possible second structure beneath the first. A wavelet processed section using a symmetric simple waveform yields a more readily interpretable section. In this case, the sand is sufficiently thick to produce non-overlapping reflections from its top and bottom. A trough (white event) signifies the low acoustic impedance as the seismic wavelet first encounters the sand, and a peak (black) characterizes the base of the sand.

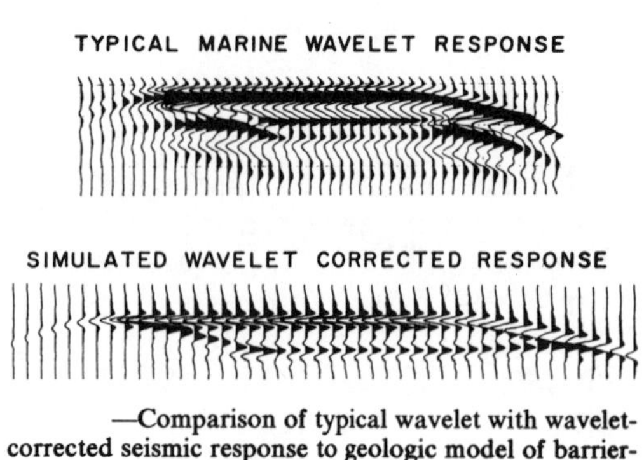

—Comparison of typical wavelet with wavelet-corrected seismic response to geologic model of barrier-bar sandstone in Figure 5.

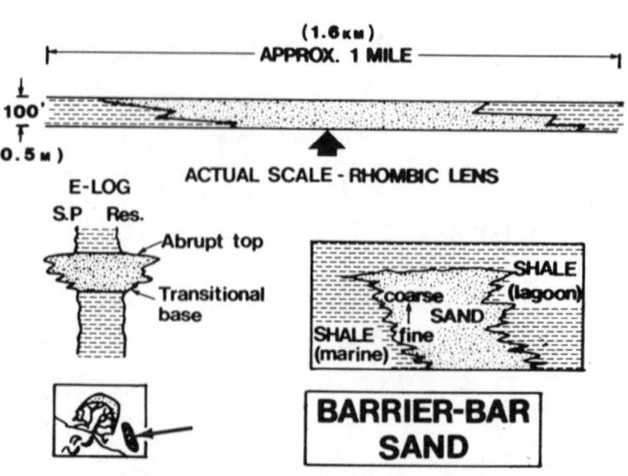

—Geologic model for a barrier-bar sandstone.

Careful consideration of these results produces still more refined interpretational clues. First, both sections show a strong content of diffraction events which is in fact one of the ususal indicators of such sands. Further, in the wavelet processed section we observe that the size of the reflection event at the sand base is less than the size of the event characterizing the sand top. On the scale of normal seismic plotting this is the principal indication of a transitionally derived reflection in contrast to one derived from an abrupt lithologic change. Here then is another interpretive clue to the identification of barrier bar sands, but one which can only practically be applied to wavelet processed seismic displays.

C. Reef Detection

Turning now to the search for reefs in Michigan we consider a model study executed by Nath (1975). The remarkable improvement in exploration success for Silurian reefs in Michigan began first with appropriate seismic data processing techniques treating the severe statics problem which had previously degraded the data quality (see in particular McClintock (1976)). Next, the preservation of seismic amplitude information for land data constituted a significant further step. Finally, the seismic characterization of the reef objective completed the interpretive picture.

As Nath's summary concludes, the reef signature is identified largely as a break in continuity for the pinnacle form. After consideration of lithologic differences in acoustic terms, we note from the ray trace display that geometrical effects of the tight curvature disperse the reef reflection response over a broad ground surface area. Hence, geometrical considerations prevail making consideration of the acoustic differences between the anhydrite, evaporite and carbonates a secondary consideration. The reef which is by definition a stratigraphic trap, has a seismic indication which seems stratigraphic in appearance yet in fact is structural in origin. Note again that the indicated seismic extent of the reef has also been much exaggerated by the same ray spreading mechanism.

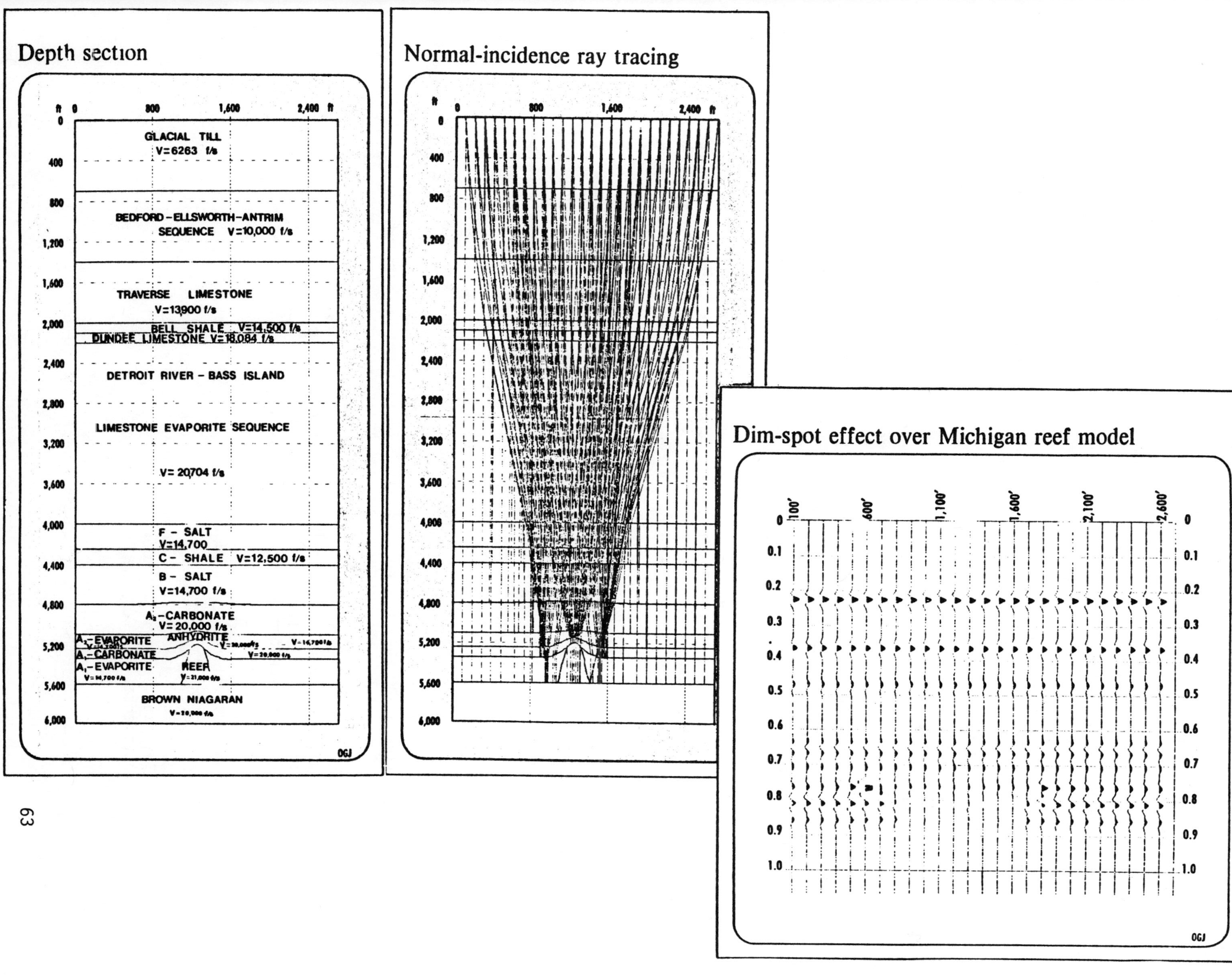
Depth section
GLACIAL TILL
V=6263 f/s
BEDFORD-ELLSWORTH-ANTRIM SEQUENCE V=10,000 f/s
TRAVERSE LIMESTONE
V=13900 f/s
BELL SHALE V=14,500 f/s
DUNDEE LIMESTONE V=18,084 f/s
DETROIT RIVER - BASS ISLAND
LIMESTONE EVAPORITE SEQUENCE
V= 20704 f/s
F - SALT
V=14,700
C - SHALE V=12,500 f/s
B - SALT
V=14,700 f/s
A2-CARBONATE
V= 20,000 f/s
ANHYDRITE
A2-EVAPORITE
A1-CARBONATE
A1-EVAPORITE
REEF
BROWN NIAGARAN
OGJ
Normal-incidence ray tracing
Dim-spot effect over Michigan reef model
100'
600'
1,100'
1,600'
2,100'
2,600'
OGJ

Since field usage is the acid test for any technical advance, we will examine a comparison between field data and a modeled section.*

The model is of the Ray Reef in Macomb County, Michigan. The Ray Reef was discovered in 1961. The Reef is a Silurian feature, approximately 2 miles long and ½ mile wide, with an axis in a NE/SW direction. Figure 14a is an accurate geological model, cutting the reef E-W along the south line of Sec. 36, T5N, R13E. Velocity information was derived from an integrated sonic log, and densities from formation density logs. The model shows a flat interface in the reef. This is the approximate gas-liquid interface. Velocities and densities in the reef were calculated by replacing pore space in the reef material with gas and water, then adjusting the assumed velocity and density of the reef material.

Figure 14b is a raypath display from the A_2 salt which caps the reef. Notice the severe focusing near the flanks of the reef. We will soon see the effect of this focusing on a seismic section. Figure 14c shows the rays emerging from all horizons.

Figure 14d is a synthetic seismic section from the model. The normal reef indicators are present. There is thinning over the reef, due to its structure. The reef displays the usual dead zone appearance. The phase reversal at the top of the reef is due to the different reflection interference on reef, compared to host sedimentation. Notice the downward curving reflection at the right flank of the reef. This shows the raypath focusing as seen on the raypath display. The left flank has similar phase distortion. The sub-reef reflection shows discontinuities beneath the reef flanks. These are supported by the defocusing through the steep sides of the reef. The model compares favorably with a VIBROSEIS[R] line shot across the reef (Figure 14e). It is the author's opinion that much information is gained by examining the reef flanks and sub-reef reflections. In this case, this information might be lost through processing, such as automatic statics.

As final notes on the Ray Reef Model, no diffractions were used because they did not play a role of particular significance on the seismic section. Also, the character of the flat-topped reef is markedly different from the pinnacle reef considered earlier by Nath.

---

* Morrison, O.J., 1975, A Discussion of the Features of the AIMS Modeling System (TM - GeoQuest International, Ltd.) - Courtesy of Seismograph Service Inc.

R - Continental Oil Company

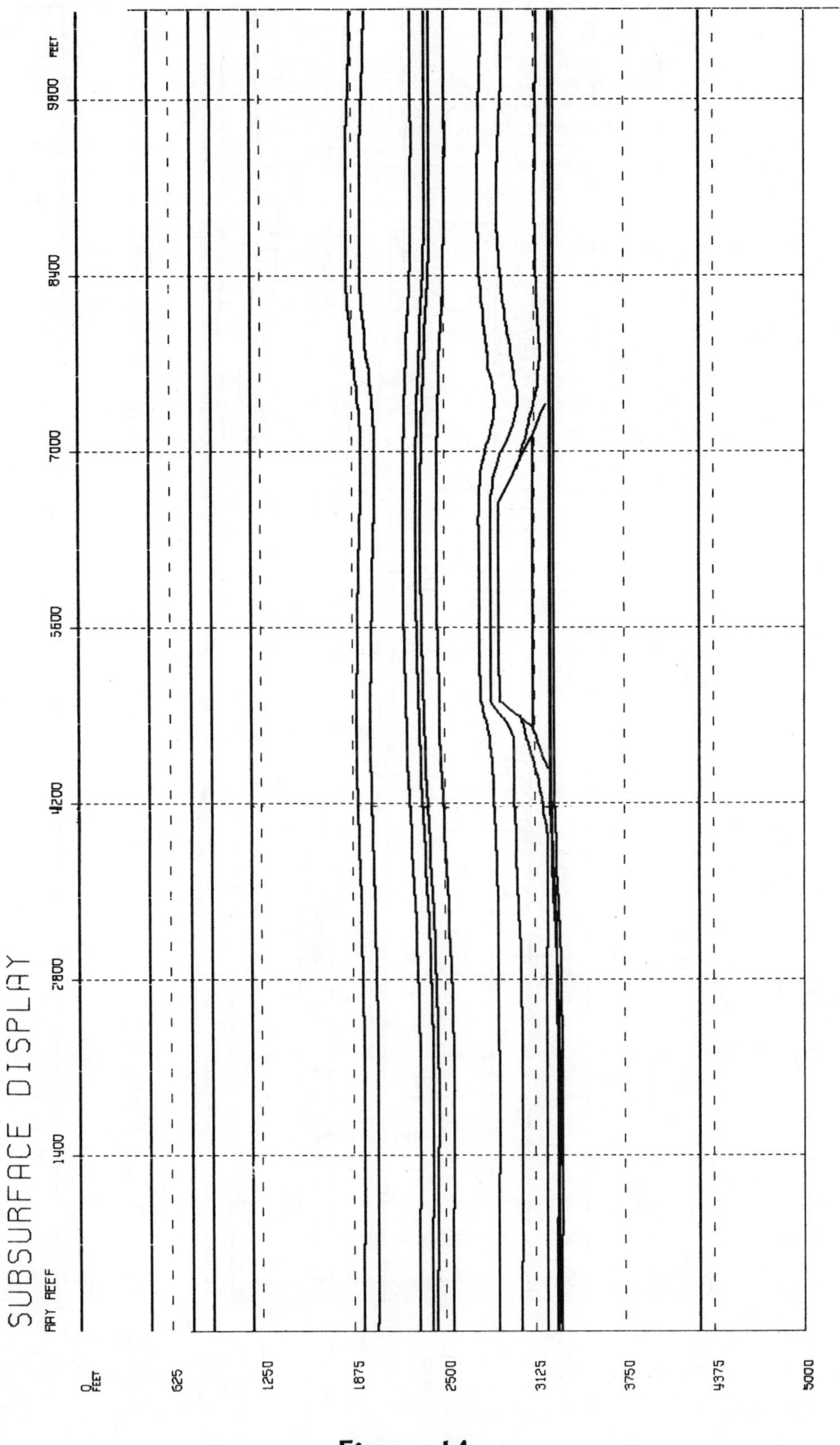
SUBSURFACE DISPLAY
RAY REEF
1400
2800
4200
5600
7000
8400
9800 FEET
0 FEET
625
1250
1875
2500
3125
3750
4375
5000

Figure 14a

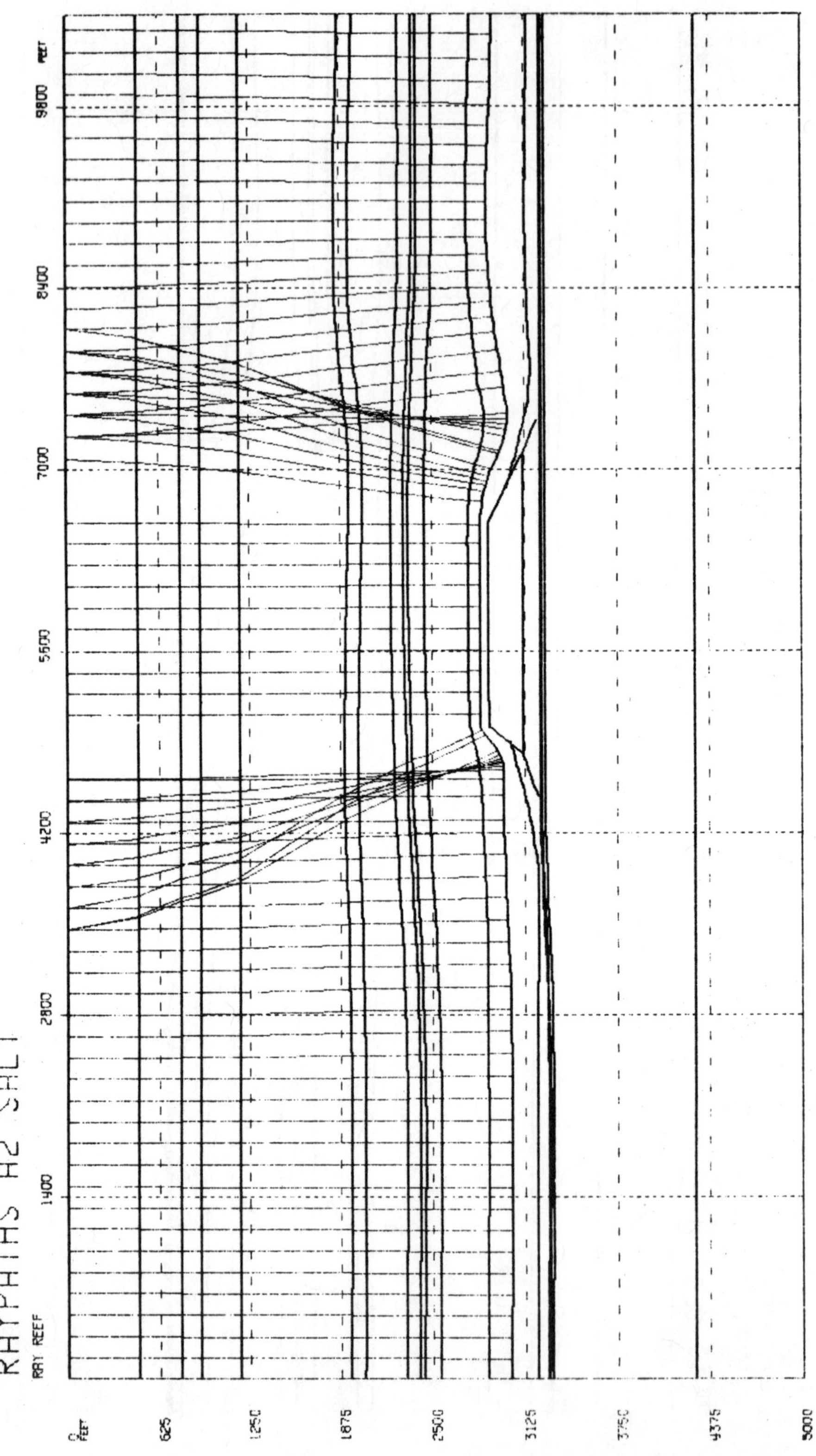
RAYPATHS A2 SALT
RAY REEF
0
FEET
625
1250
1875
2500
3125
3750
4375
5000
1400
2800
4200
5600
7000
8400
9800
FEET

Figure 14b

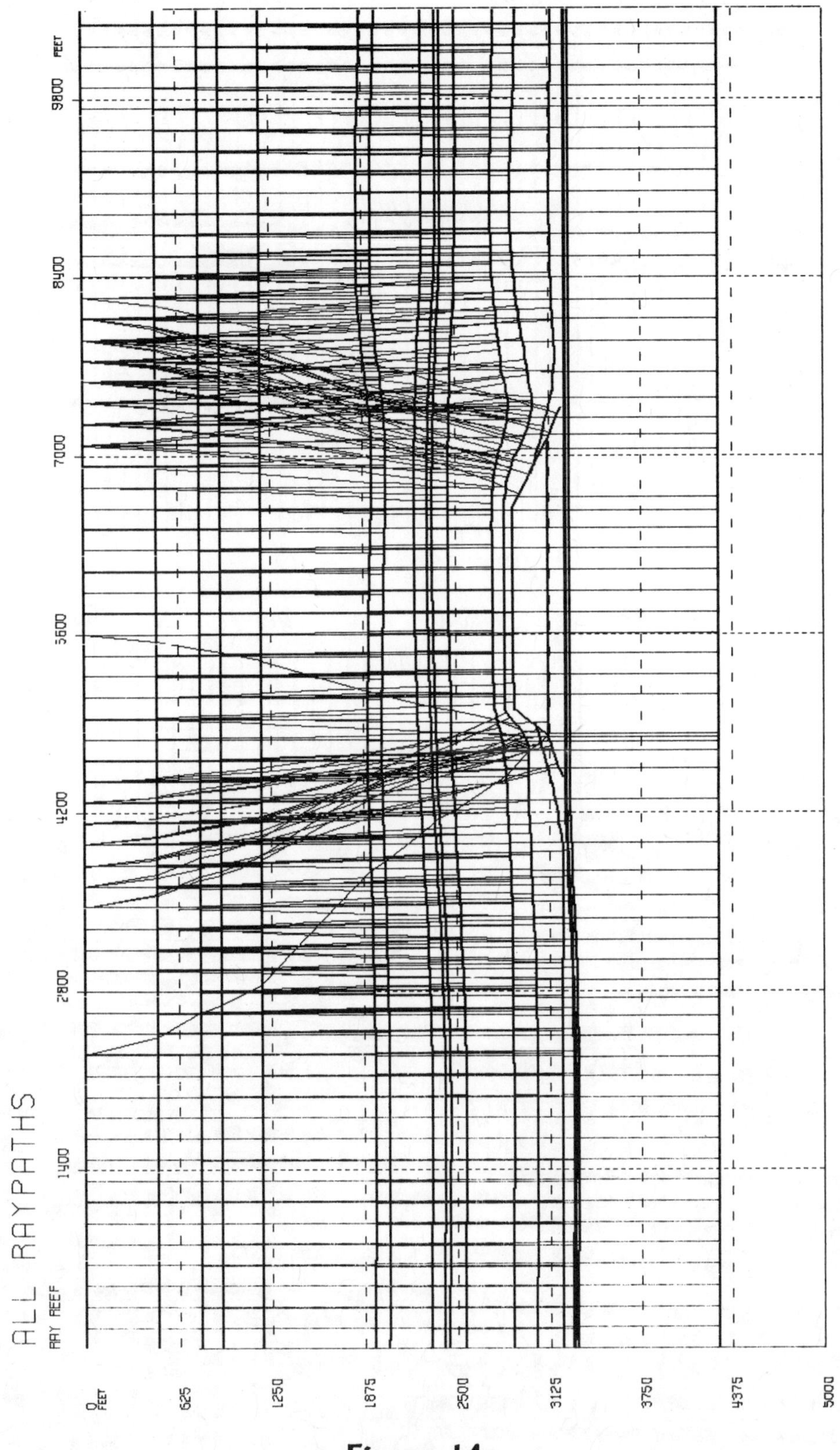
ALL RAYPATHS
RAY REEF
1400
2800
4200
5600
7000
8400
9800 FEET
0 FEET
625
1250
1875
2500
3125
3750
4375
5000

Figure 14c

Figure 14d

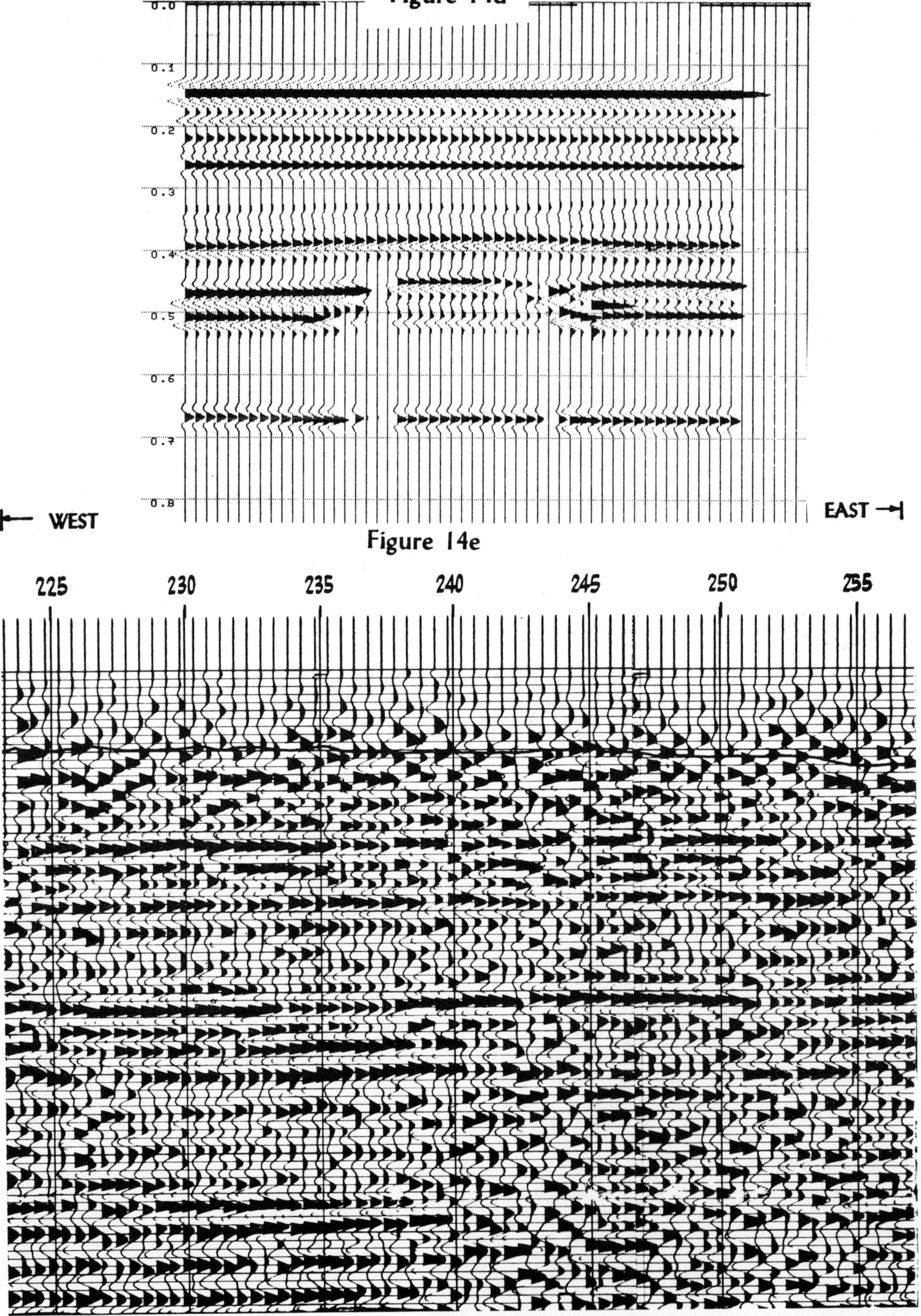

Figure 14e

D. Stratigraphic Correlation

Another aspect of the role of geoseismic modeling can be understood by reviewing an illustrative use from an actual problem in the North Sea where we are concerned with understanding the stratigraphy between two wells. A seismic line was shot connecting two wells. The seismic information showed very poor signal-to-noise ratio typical of the area, and therefore conventional interpretation of the seismic data was ambiguous at best. The propagating wavelet information was extracted from previous data shot by the same acquisition boat in the same area. This wavelet was used to wavelet process the data and was also used in modeling.

The next Figure called Depth Section - Model A displays formation parameters determined from the logs of these two wells which are joined by an interpreted subsurface termed Model A. At Well A, a shale with velocity of 2134 meters/second overlies a higher velocity shale (2438 meters/second). This high-velocity shale has some sand stringers below a depth of 2,000 meters. Well B, some 3100 meters away from Well A, has quite a different lithology. The low-velocity shale is underlain by a 112 meter sand unit, the upper 30 meters of which is gas filled. The high-velocity shale is missing in this well.

Geologic model prepared from data available from two well penetrations

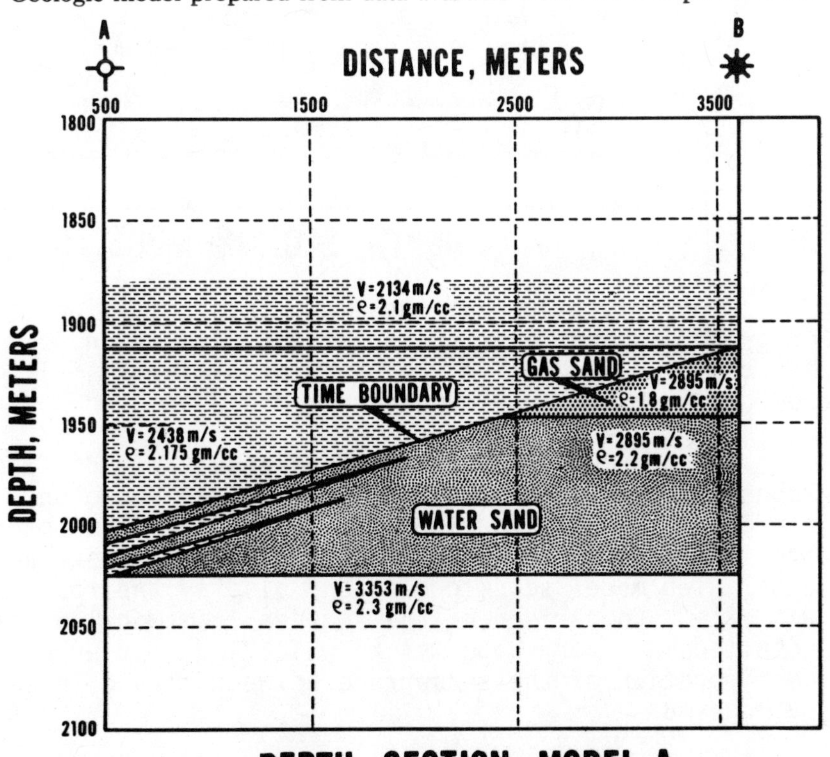

DEPTH SECTION - MODEL A

These same logs were interpreted by different groups of geologists and engineers and the interpretations were quite different, even though both honored the observed data at the well penetrations. Another possible geologic model is shown in Depth Section - Model B. The interpretation of the seismic response and the geology between these wells boils down to resolving the stratigraphic nature of the upper boundary of the producing sand unit. Model A represents the case where the upper boundary of the sand is also a time boundary. Model B on the other hand, shows an abrupt facies change between the two wells where there is an interfingering of sand and shale and the time boundary now is assumed to be horizontal gradational acoustic interface between the two shales.

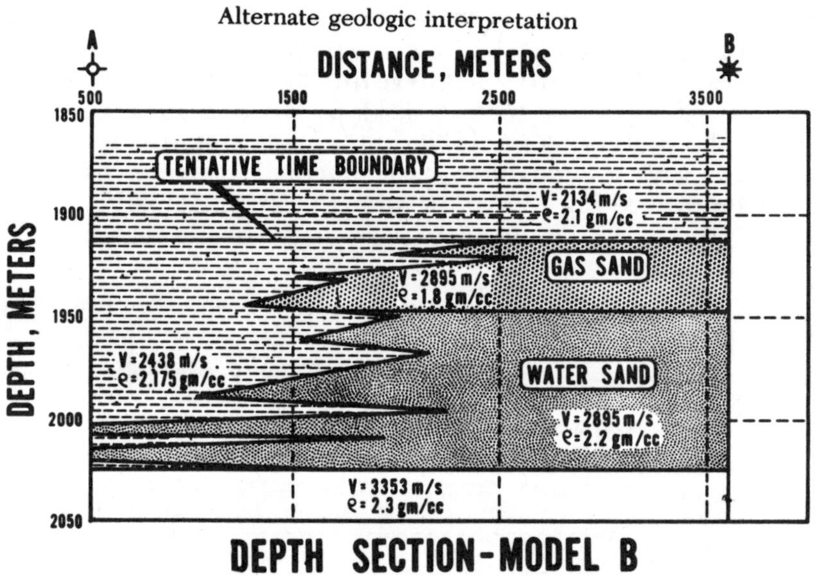

The implications of these two interpretations are quite clear and of great importance in the reservoir definition. If the situation is as predicted by Model A, we have a relatively small field. Model B, however, represents a much larger field.

The two models may well yield resolvably different seismic responses in the zone of interest. Seismic response simulation was carried out. Both models were sampled every 50 meters. The reflectivity series were viewed with the basic wavelet extracted from the seismic data. Two model seismic sections display the results of the modeling exercise. Comparison of the section from Model B with the actual data resulted in a very good match, which led to the conclusion that the configuration of the subsurface between the wells is better represented by Model B.

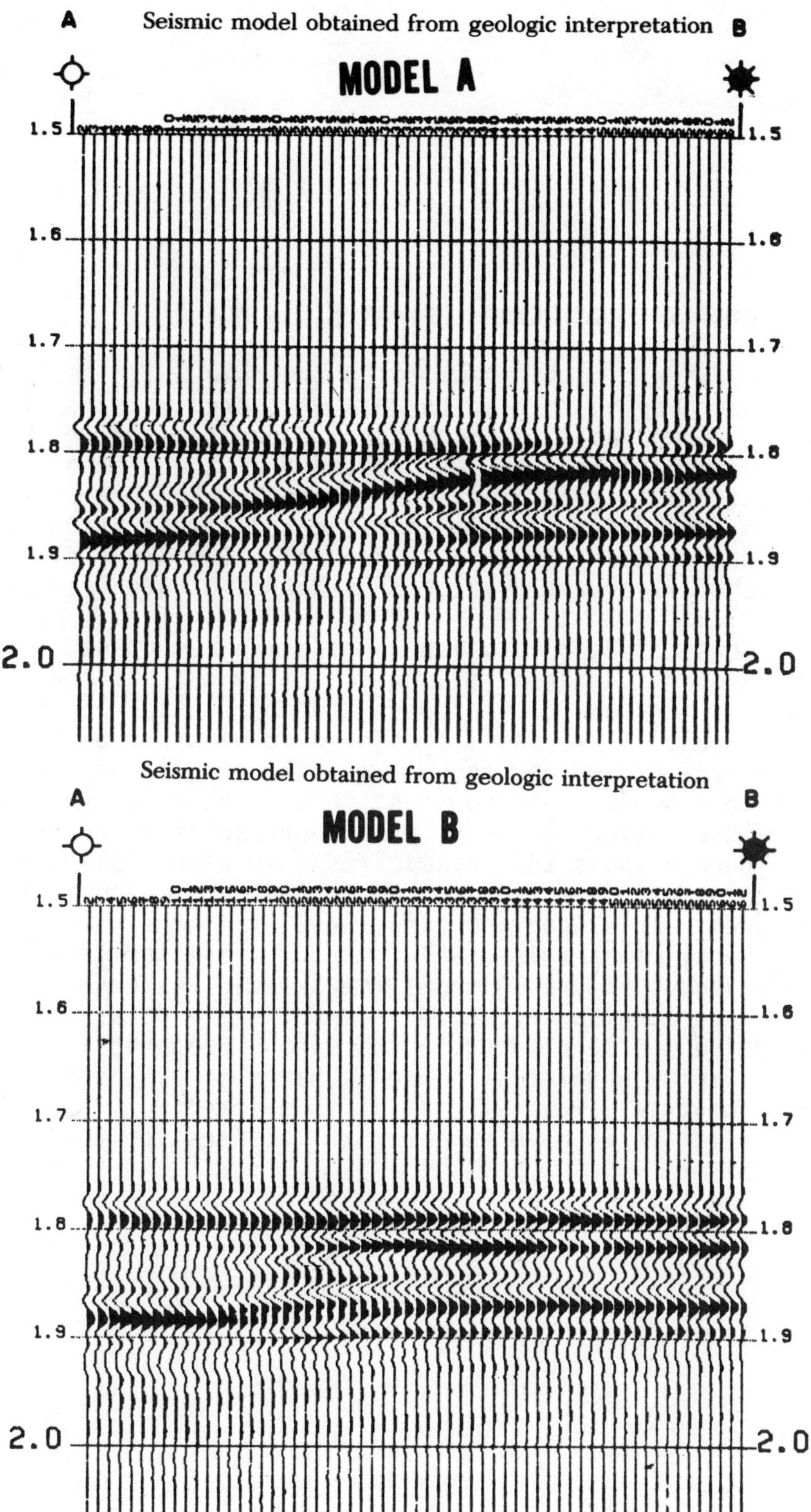

This example points out that modeling is useful and even a necessary step with potentially great economic consequence for drawing interpretive conclusions from seismic data. We have also found that the practice of modeling sections aids greatly in training young geophysicists and interpreters. Many questionable interpretations now may be proven by the use of geoseismic modeling before they are proven by the drill.

E. Vertical Resolution - Brent Field Model

In an earlier discussion we noted the role which geoseismic modeling might play in resolution when we looked at sand bodies whose extension was comparable to the Fresnel Zone diameter or less. The matter of resolution will now be treated from the viewpoint of a model subsurface of practical exploration interest. Further, the role of seismic data processing is also investigated illustrating the multi-faceted nature of model studies in general. The Brent field text and figures which follow were developed by Nath and Patch (1978).

The process of deconvolution makes a minimum phase assumption about the wavelet. It then follows logically that if the basic wavelet has a true minimum-phase property, deconvolution would produce a satisfactory result. We shall present a model study performed on a Brent-field type of geology and examine the results of conventional deconvolution procedures in contrast to wavelet processing. Figure 8 illustrates the subsurface model with representative acoustic parameters. Figure 9 shows the seismic response with a sleeve exploder wavelet, which was extracted from North Sea data. Some random noise has been added. This long and ringy wavelet is the Wavelet shown in Figure 9. Recall that this wavelet is quite similar to its minimum phase equivalent. Figure 10 shows the seismic response with a tuned air-gun signature, also extracted from North Sea data. A similar amount of random noise has been added to the data. Compared to these two sections with complex wavelets, Figure 11 has a simple, short, zero-phase band pass symmetrical wavelet. It is immediately obvious that stratigraphic correlation is much easier with this section than with the previous two.

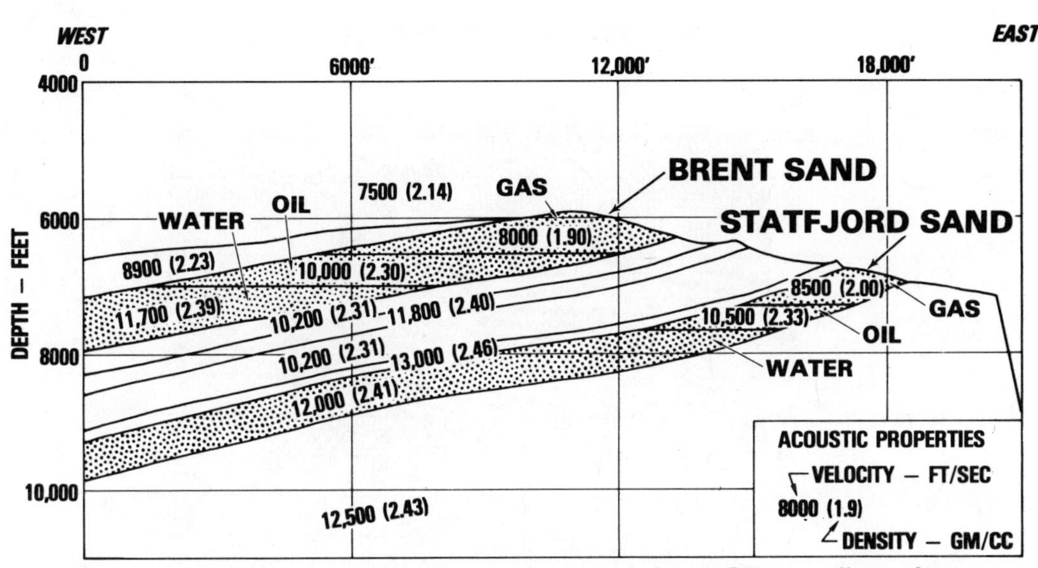

Fig. 8 - Acoustic depth model of Brent Field in North Sea.

Standard deconvolution techniques would normally be applied in the seismic sections with the marine wavelets. Figures 12 and 13 illustrate the results of applying a spiking deconvolution operator to the sections with sleeve exploder and air-gun signatures, respectively. This operation seems to have yielded reasonably satisfactory results in the case of the seismic data with the minimum-phase type sleeve exploder signature, but has produced a poor result in the case of the air-gun signature.

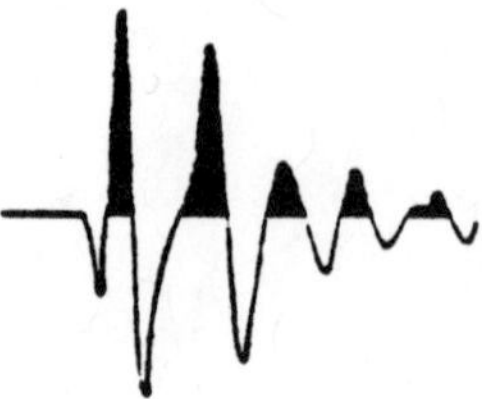

SLEEVE EXPLODER WAVELET FROM N. SEA DATA

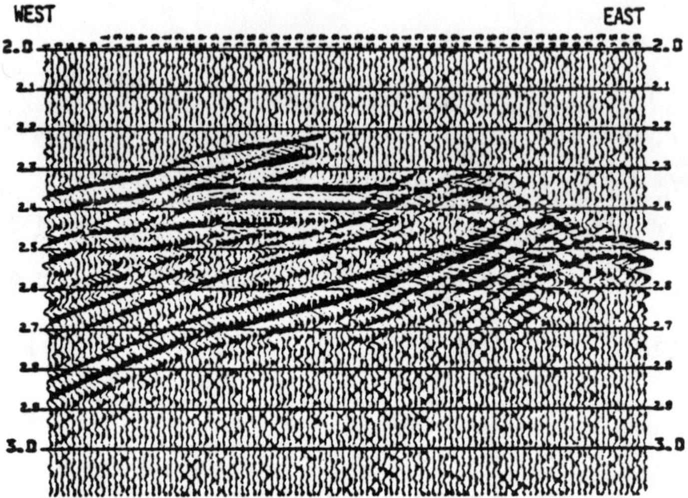

FIG. 9 - BRENT FIELD TIME SECTION WITH SLEEVE EXPLODER WAVELET (NOISE ADDED).

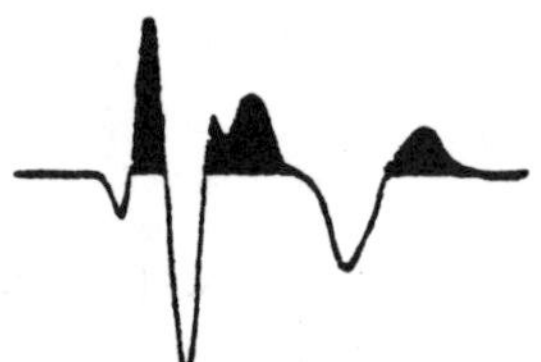

TUNED AIR GUN WAVELET FROM N. SEA DATA

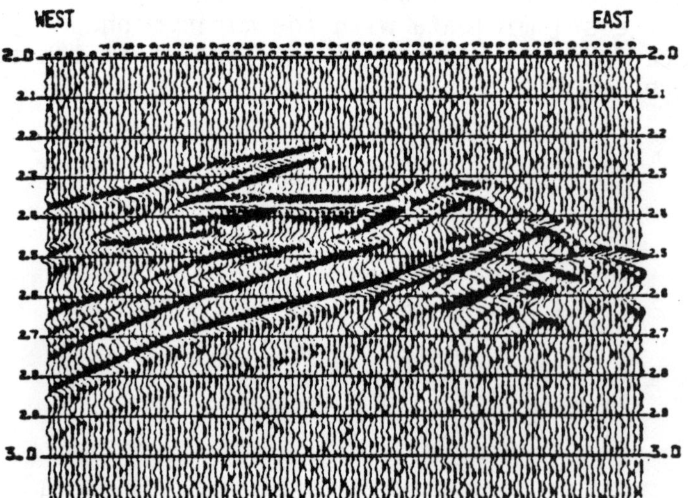

FIG. 10 - BRENT FIELD TIME SECTION WITH AIR GUN WAVELET (NOISE ADDED).

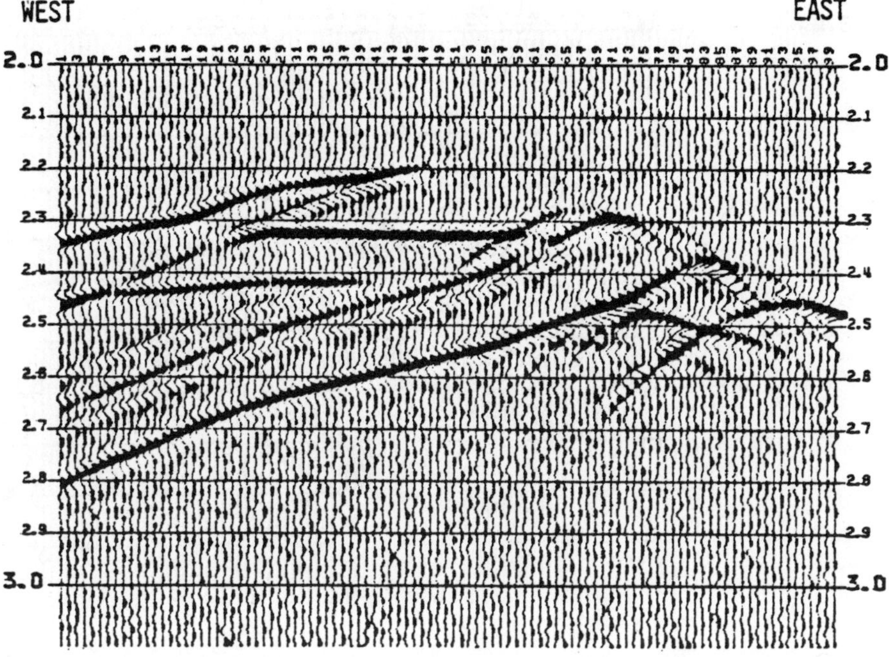

Fig. 11 - Brent Field time section with 10-40 hertz zero phase Butterworth bandpass wavelet (noise added).

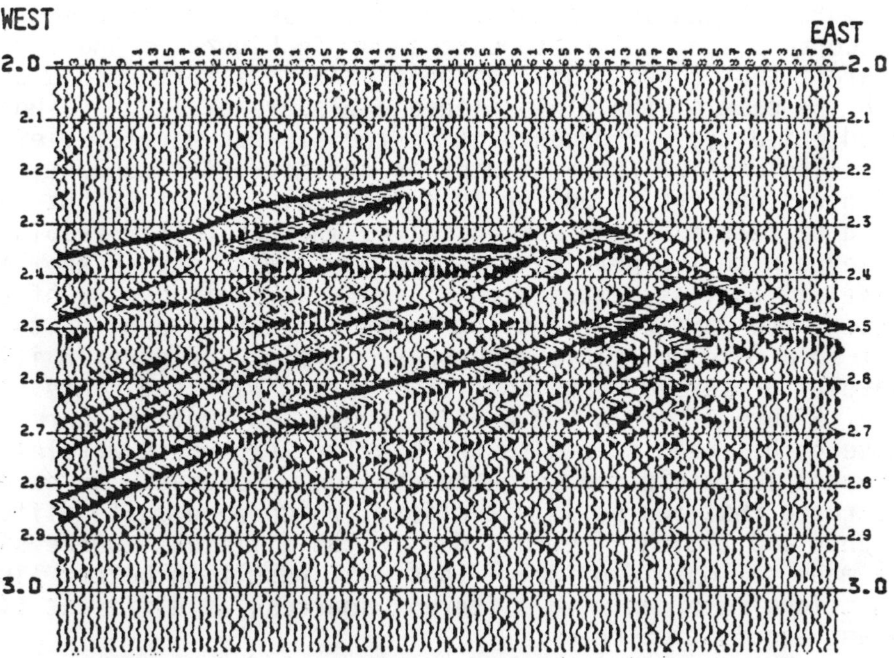

Fig. 12 - Brent Field time section with sleeve exploder wavelet (spiking deconvolution applied).

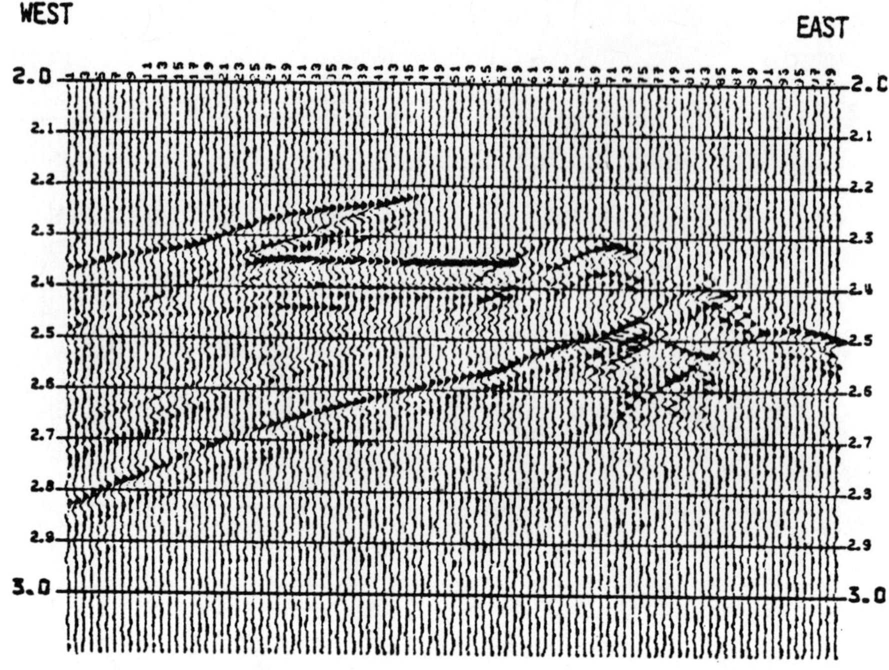

FIG. 13 - BRENT FIELD TIME SECTION WITH AIR GUN WAVELET (SPIKING DECONVOLUTION APPLIED).

Figures 14 and 15 are displays of the seismic responses with predictive deconvolution applied to data shown in Figures 9 and 10. Again it is apparent that conventional deconvolution has performed a better result in the case of the ringy sleeve exploder wavelet. In both spiking and predictive type of deconvolution the air-gun wavelet yielded a poor result. The minimum phase assumption necessary for conventional deconvolution was not satisfied for this particular wavelet. At this point it would be reasonable to ask ourselves whether better resolution could be achieved with the new found technology. With proper knowledge of the wavelet, that is, extraction of these wavelets by one of the methods discussed, the wavelet processing technique can be applied to shape this wavelet to a more desirable zero-phase wavelet. Figure 16 illustrates the wavelet processed version of the section with the ringy sleeve exploder. Figure 17 shows the wavelet processed section where the air-gun wavelet has been shaped to the simpler zero-phase wavelet. The improvement in resolution is quite obvious. Moreover, we have also demonstrated that two seismic sections, which were quite different in appearance and timing of events due to two different source signatures, can be made to look identical. This will result in better line ties with sections recorded with different sources, at different times by different acquisition crews.

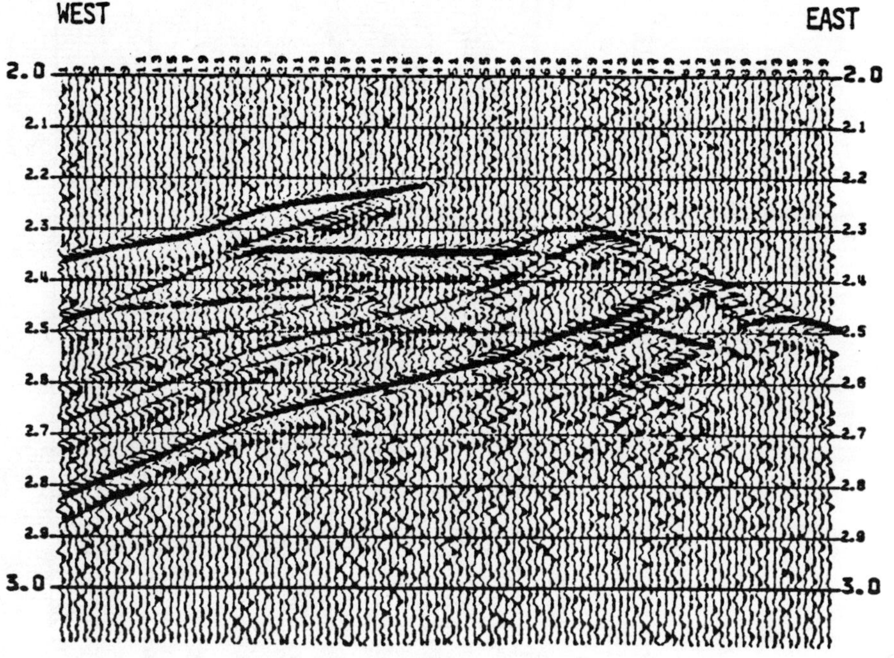

FIG. 14 - BRENT FIELD TIME SECTION WITH SLEEVE EXPLODER WAVELET (PREDICTIVE DECONVOLUTION APPLIED).

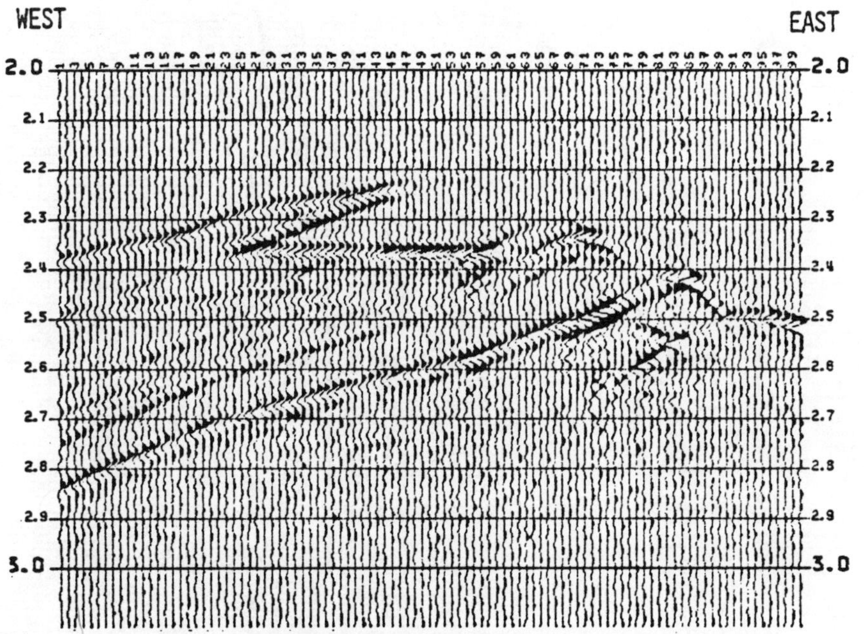

FIG. 15 - BRENT FIELD TIME SECTION WITH AIR GUN WAVELET (PREDICTIVE DECONVOLUTION APPLIED).

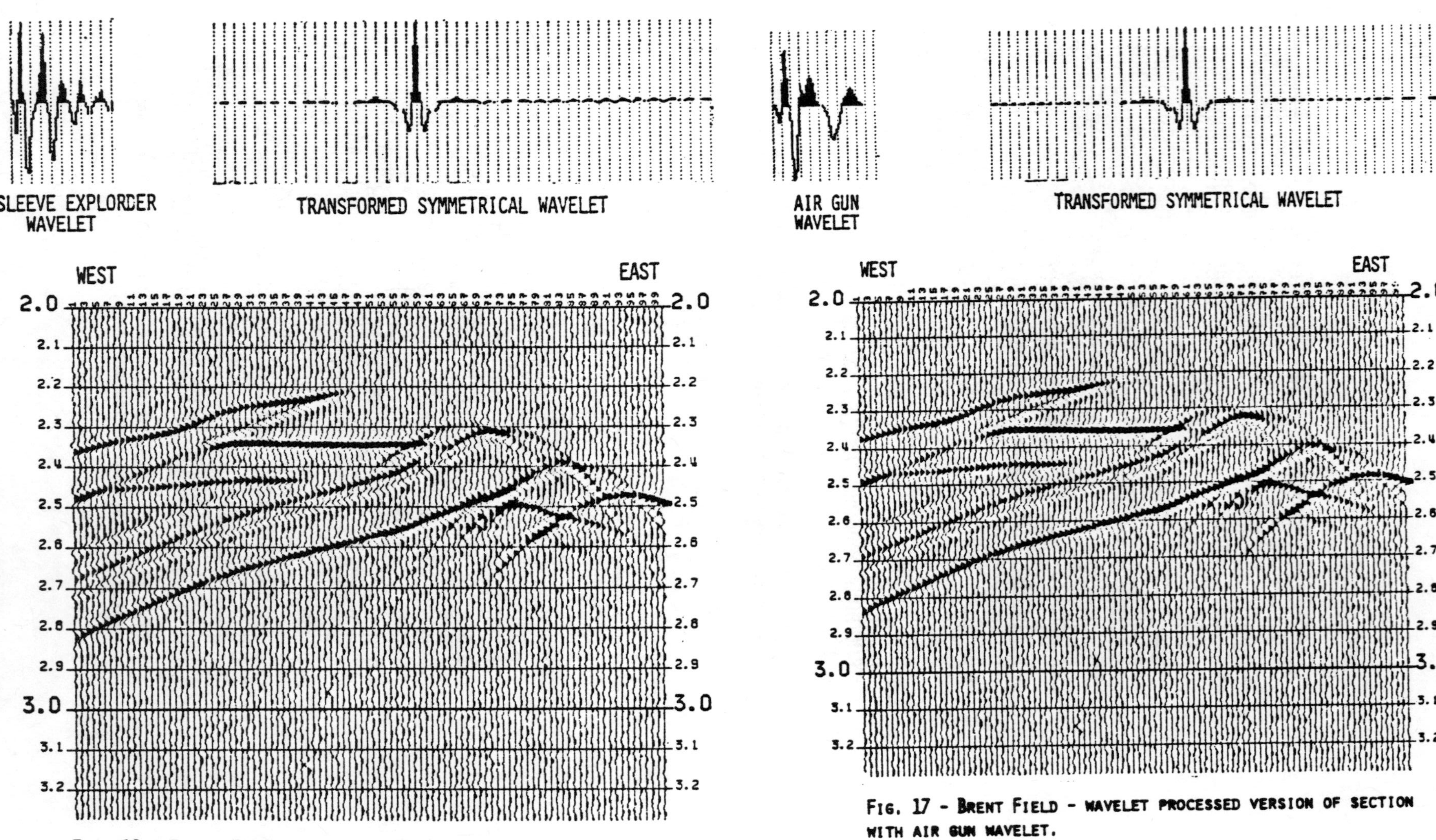

Fig. 16 - Brent Field - wavelet processed version of section with sleeve exploder wavelet.

Fig. 17 - Brent Field - wavelet processed version of section with air gun wavelet.

E. Lateral Resolution - A Study of a River Mouth Sand Bar.

The criterion of delineating boundaries on the basis of recognizing one-half reflection amplitudes on the beginning of diffraction amplitudes has been used quite extensively during seismic interpretation. How well can this concept, together with the amplitude information about Fresnel zones (as discussed previously), help geophysicists engaged in stratigraphic interpretation? We shall attempt to answer this question and a few more by performing a three dimensional modeling study of a river mouth sand bar.

The figure on this page is a map view of the amorphous sand body. The depth of burial to this structure is 5,000 feet and the relief is so small as to be beyond vertical resolution of the seismic system. Five north-south seismic lines and five east-west trending lines are designed on a 1,000 foot grid over this feature. From the Fresnel zone experiments, it appears that the edge reflections should be one-half the amplitude of a normal incidence reflection, so that we can utilize this criterion to pick the edge of the sand body. This amplitude rule was used on the wavelet sections shown in the next two pages.

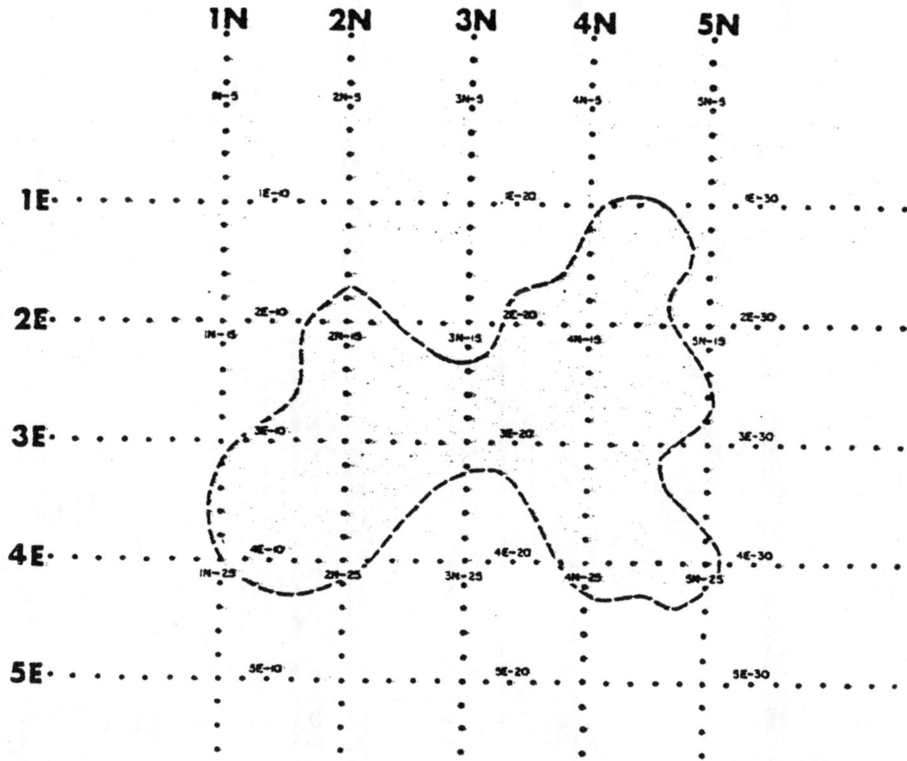

Base maps of a flat small sand body which is buried at 5,000 feet. Seismic lines are 1,000 feet apart.

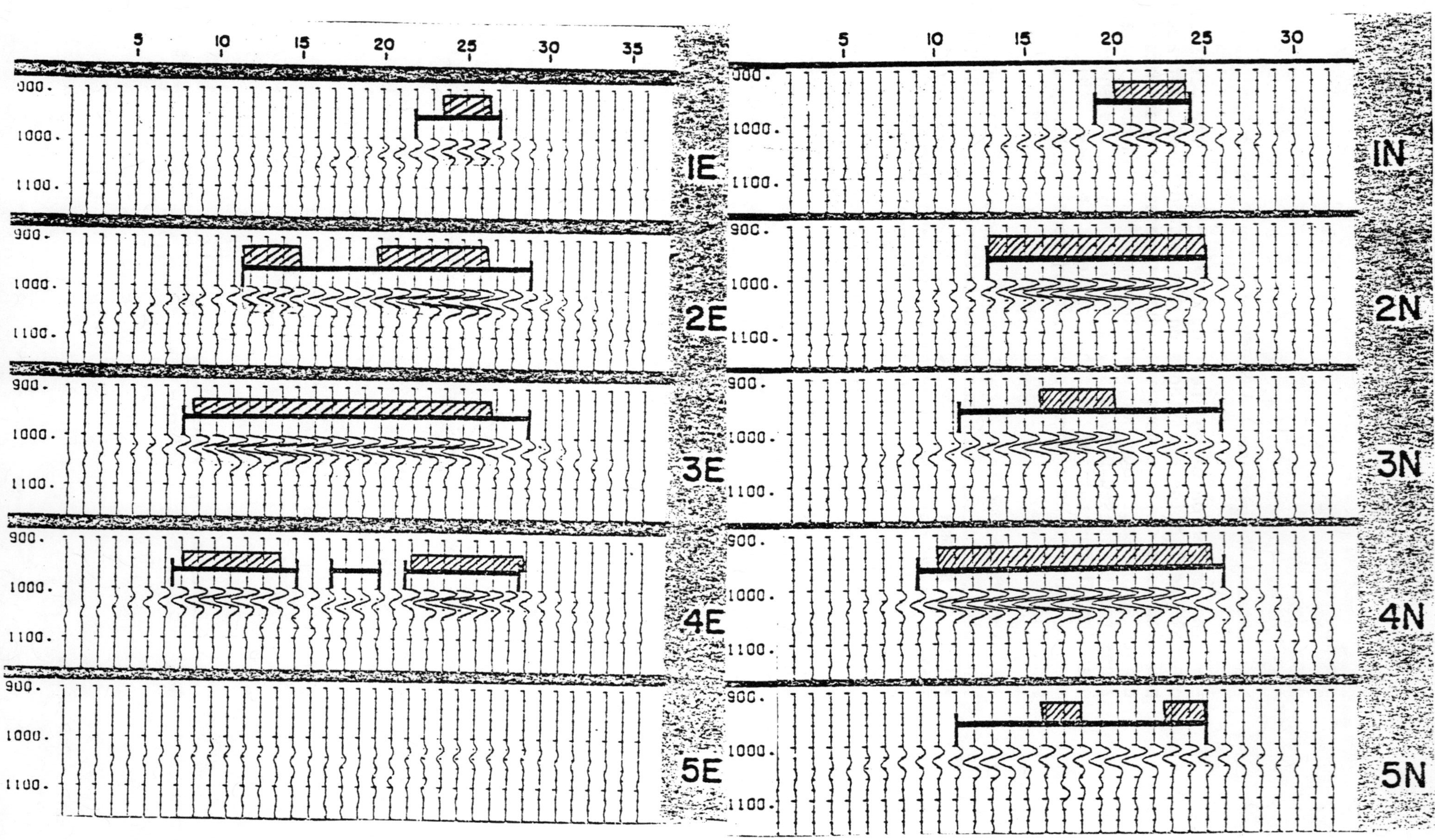
1E
2E
3E
4E
5E
1N
2N
3N
4N
5N
5
10
15
20
25
30
35
900.
1000.
1100.

In the wavelet time sections in the last two figures, the shaded stripe represents the extent of the geologic boundary, while the black line represents the interpreted areal extent using the half amplitude rule. On Line 2E, the edge of the field at the 3N intersection is not seen and therefore, the field is overestimated. This is a problem due to a concave edge as seen previously by the Fresnel hole studies. Diffraction healing has caused the anomalous events, which have been interpreted as reflection from a continuous body.

In a similar fashion, Line 4E overestimates the extent of the field. The energy buildup in the middle of 4E is due to a positive branch of diffractions from the inside edges of the two reflectors. That is, in the middle of Line 4E, the erroneous sand bar was predicted when the shotpoints approached the focal line of the concave edge. Unfortunately, the small time delay where the erroneous sand was predicted on Line 4E is approximately .004 seconds and thus would normally be ignored by associating it with either structure or a velocity anomaly over the field. In a similar fashion, the diffraction events off the edge are too small to do an effective diffraction interpretation.

Line 3N goes through the neck of the field and the resulting wavelet section grossly overestimates the field extent. To get an idea of the magnitude of this error, the one-half amplitude rule overestimates by 1,000 feet on the north side and by 1,200 feet on the south side. An outline of the estimated field, as predicted by the half-amplitude rule from the wavelet time section, is shown in a map view on the following page.

The dotted area, along with the original shaded area, is the seismic estimate and this happens to be 40% too high. All the overestimation occurs by concave edges. Of course, this error can be significantly reduced if a denser seismic grid was used and the concave focusing phenomena are recognized during interpretation.

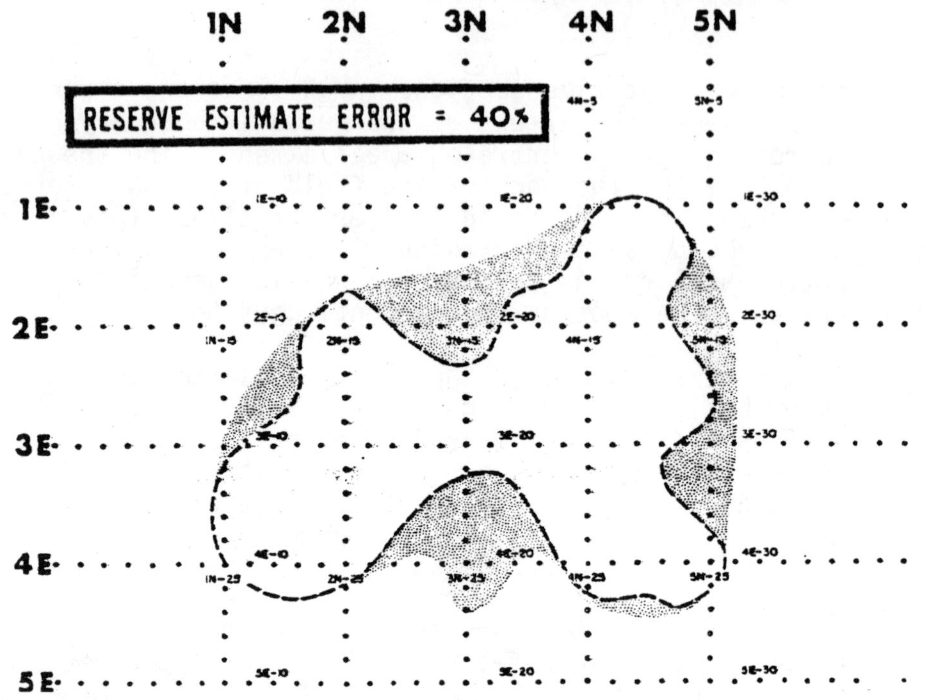

Base map with seismic interpretation of field outline using ½ amplitude rule on seismic sections.

The geoseismic model studies noted both the basic power of the tool and a number of important interpretational considerations. Sand identification in terms of the barrier-bar also gave insights into use of the waveform for a most refined application. In considering reefs, both structural considerations and the stratigraphic effects which change seismic amplitudes were observed in complex interaction. The stratigraphic correlation study modeled after a North Sea case illustrated mainly a procedural approach. Finally, the pointbar sand lateral resolution study, while a model, really comments on a fundamental resolution matter. In fact, this study in concert with the box-car reef study noted earlier explains why "corners" never appear on seismic time maps having either concave or convex nature in the horizontal plane.

While we have divided this work into sections, the overlap and interplay of the technology we consider will continue. This will be quite evident in the next section where we consider the qualitative nature of seismic stratigraphic correlations, but mainly via seismic model studies.

VI. Qualitative Stratigraphic Correlations

A. Lateral Transitioning

With the tools available, we may profitably consider case studies both simulated and real to develop guidelines and procedures for stratigraphic studies. As a first example, a complex stratigraphic situation described by four lithology logs will be considered. The particular case is termed the sand-shale interfingering model.

As all geologists know lithologic boundaries are not always mirror smooth in the subsurface. Depositional characteristics of interbedding gradiation can cause a diffuse boundary in depth. This lack of a well-defined boundary will understandably influence the seismic reflections as was already noted. Additionally, there will be a lateral averaging over the Fresnel zone which usually encompasses a region of between 300-1500 feet in diameter.

When a diffuse boundary is involved we have noted that to first order there is some weakening of the reflection response. Also, there will be some measure of weakening in lateral continuity. Further, we may anticipate that the seismic response may differ from a response predicted from well log measurements via simple seismogram synthesis even if a common seismic waveform is used. The log measures the vertical lithologic sequence using a few inches of lateral penetration at the most. Seismic waves average several hundred feet laterally and thus reveal different properties from the log about the boundary.

This model attempts to portray some of the effects of a diffuse boundary. A sand of approximately 100 feet thickness is embedded in a shale. The sand base is very uniform over a large area, but the sand top is locally interfingered with the overlying shale. Four holes are shown at locations A through D. The lithology logs for these holes are shown in the second figure. Working from these alone, the sands are correlated as shown in the third figure. In the vicinity of the logs, the sand profile developed is quite close to that in the original model.

A seismic wave theory response to the complete sand unit is shown in the following figure. The upper version uses a seismic wavelet that is symmetrical and polarized such that a positive reflection (transition from a "soft" rock to a "hard" rock) will be displayed as a black or right hand deflection. This is an ideal wavelet for visual detection of detail stratigraphy and acoustic lithology. The lower seismic response is the same model using a wavelet actually found for a marine seismic line. The wavelet is not symmetrical or regular in any sense. The following figures compare both wavelets A and B for the seismic response to the synthetic seismogram derived from the logs using the same wavelets. The synthetic six-trace set is shown adjacent to the well location in each case shown.

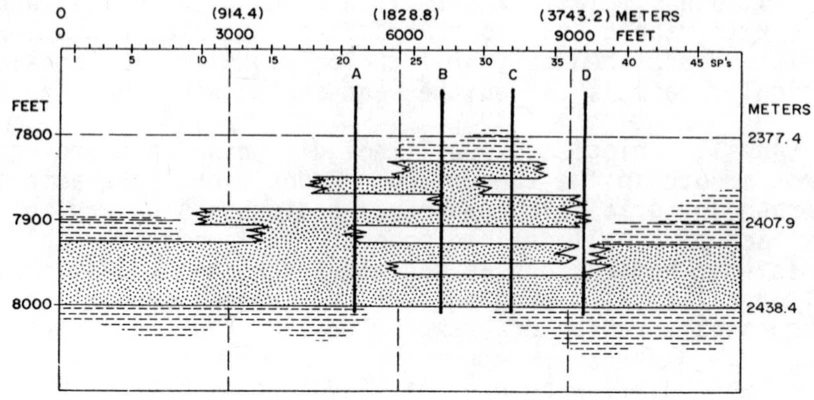

—Sandstone-shale interfingering model illustrating lateral lithologic transition.

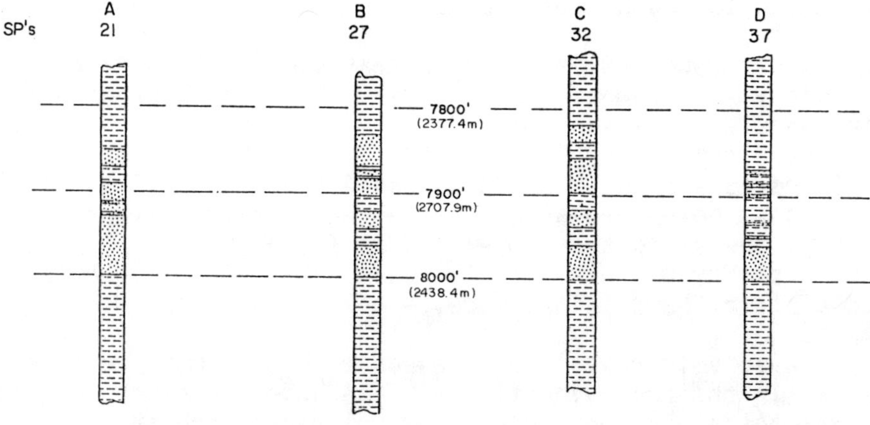

—Lithologic logs for four wells of sandstone-shale seismic section.

LITHOLOGIC LOGS FOR FOUR WELLS
ON SEISMIC SECTION
SHOWING CORRELATION OF SANDS

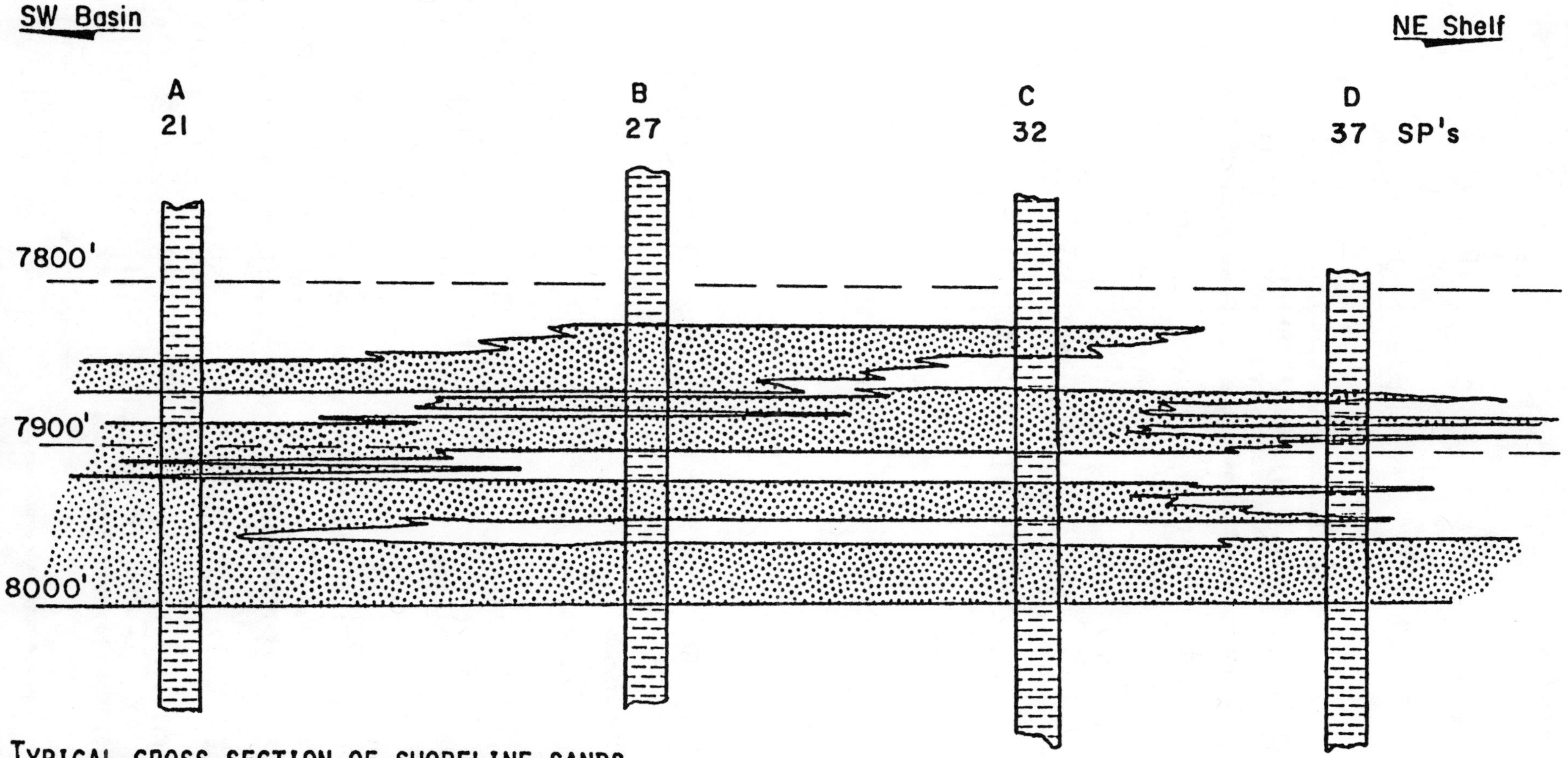

SW Basin

NE Shelf

A 21

B 27

C 32

D 37 SP's

7800'

7900'

8000'

TYPICAL CROSS SECTION OF SHORELINE SANDS IN MORROW FORMATION (LOWER PENNSYLVANIAN) ON SHIELD OF ANADARKO BASIN.

MAY ALSO REPRESENT REVERSE RELATIONSHIP OF GULF COAST TERTIARY SANDS WITH MARINE SIDE TO RIGHT (OFFSHORE) AND LAGOONAL SIDE TO LEFT.
NOTE THE RHOMBIC-SHAPED BARRIER BAR IN UPPER SAND.

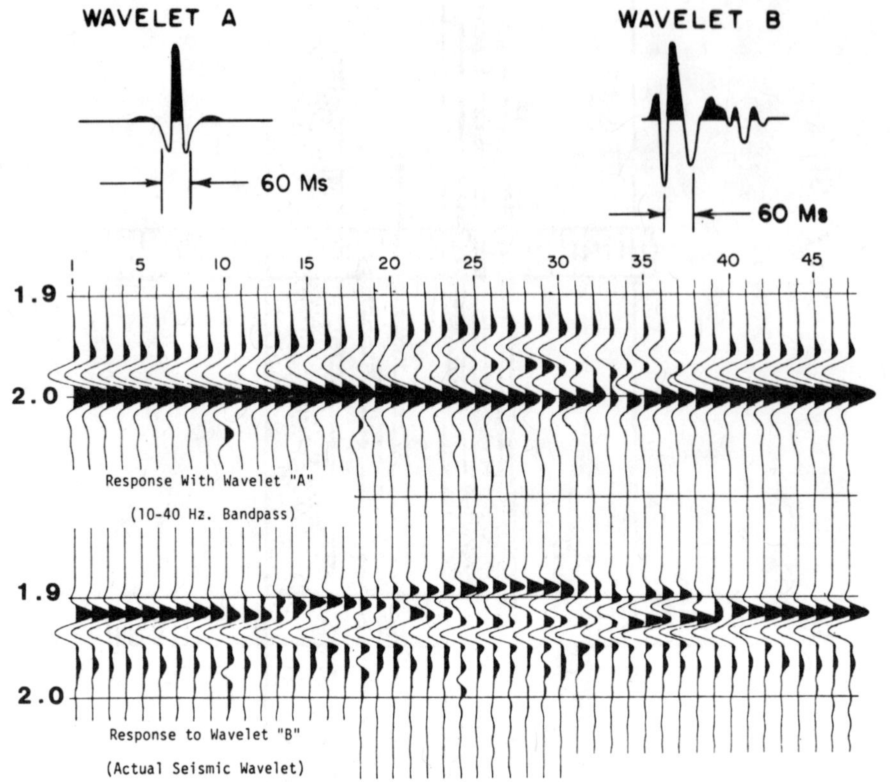

—Sandstone-shale interfingering-model seismic response.

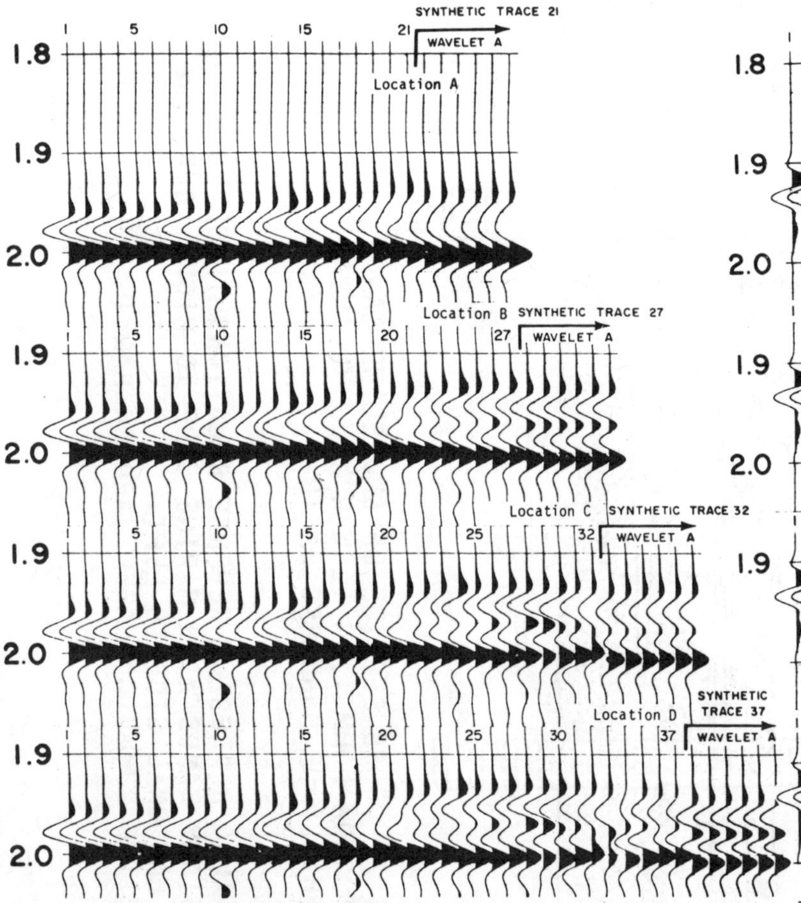

—Comparison of seismic response with wavelet A to synthetic seismogram using wavelet A.

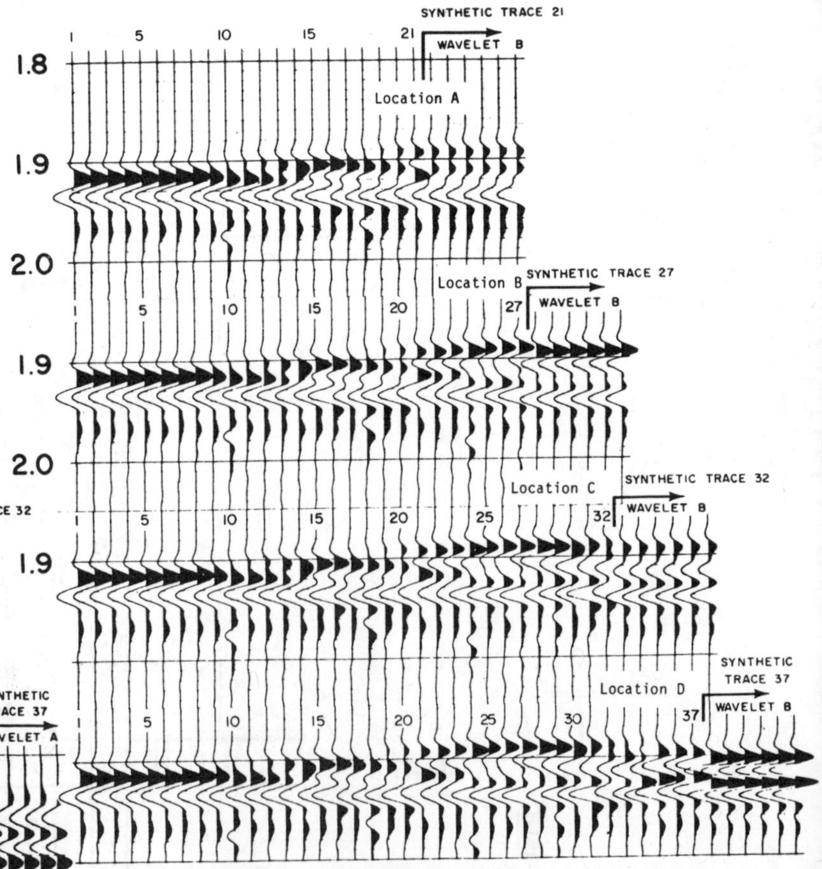

—Comparison of seismic response with wavelet B to synthetic seismogram using wavelet B.

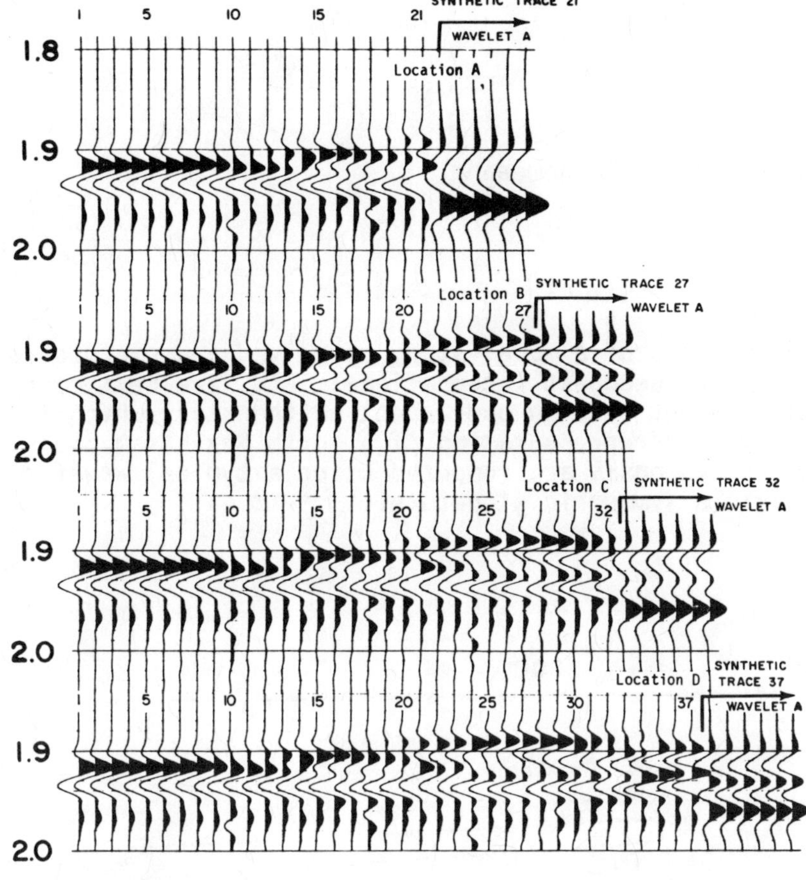

—Comparison of seismic response with wavelet B to synthetic seismogram using wavelet A.

Our observations from this study offer important guidelines for using simple synthetic seismograms. First, as we have stressed several times, it is essentail that the seismic waveform be common to the synthetic and the actual data. Next we understand that the synthetic seismogram and the actual data should correlate only to the extent that the sequence of the log extrapolates laterally for at least a Fresnel zone distance. It is particularly interesting to see how the application of more advanced tools does not rule out the use of the more elementary ones, but rather instructs us in their proper and appropriate use.

B. Pseudostratigraphic Phenomena

Abridged results of a model study conceived and executed by E. Poggiagliolmi of GeoQuest International, Ltd. are next considered. Here we shall comment only on one aspect of the study which is based on a schematic representation of a North Sea horst block. The wave theory responses are computed using a typical waveform and the more desirable symmetric signature.

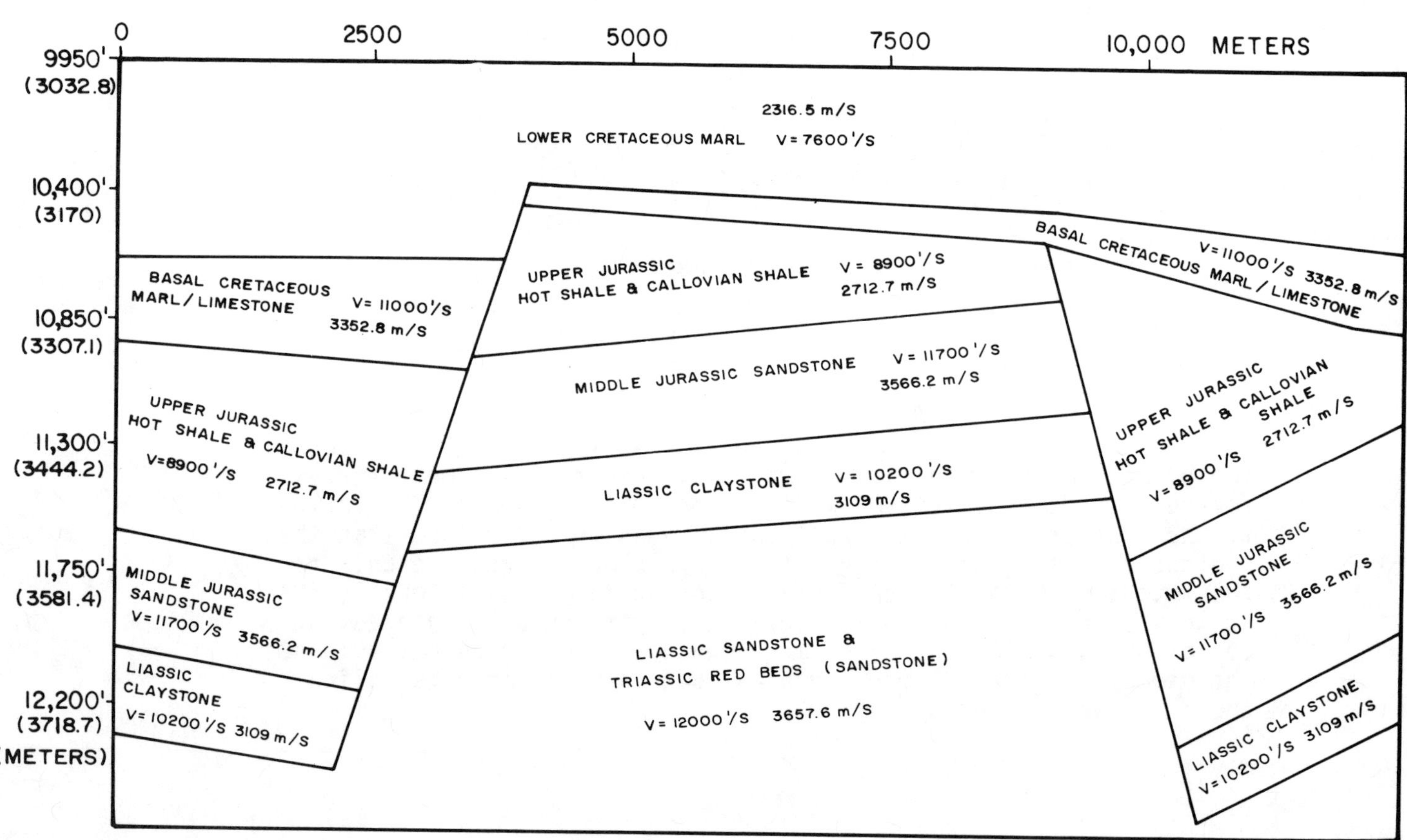

—North Sea model, geology and seismic parameters.

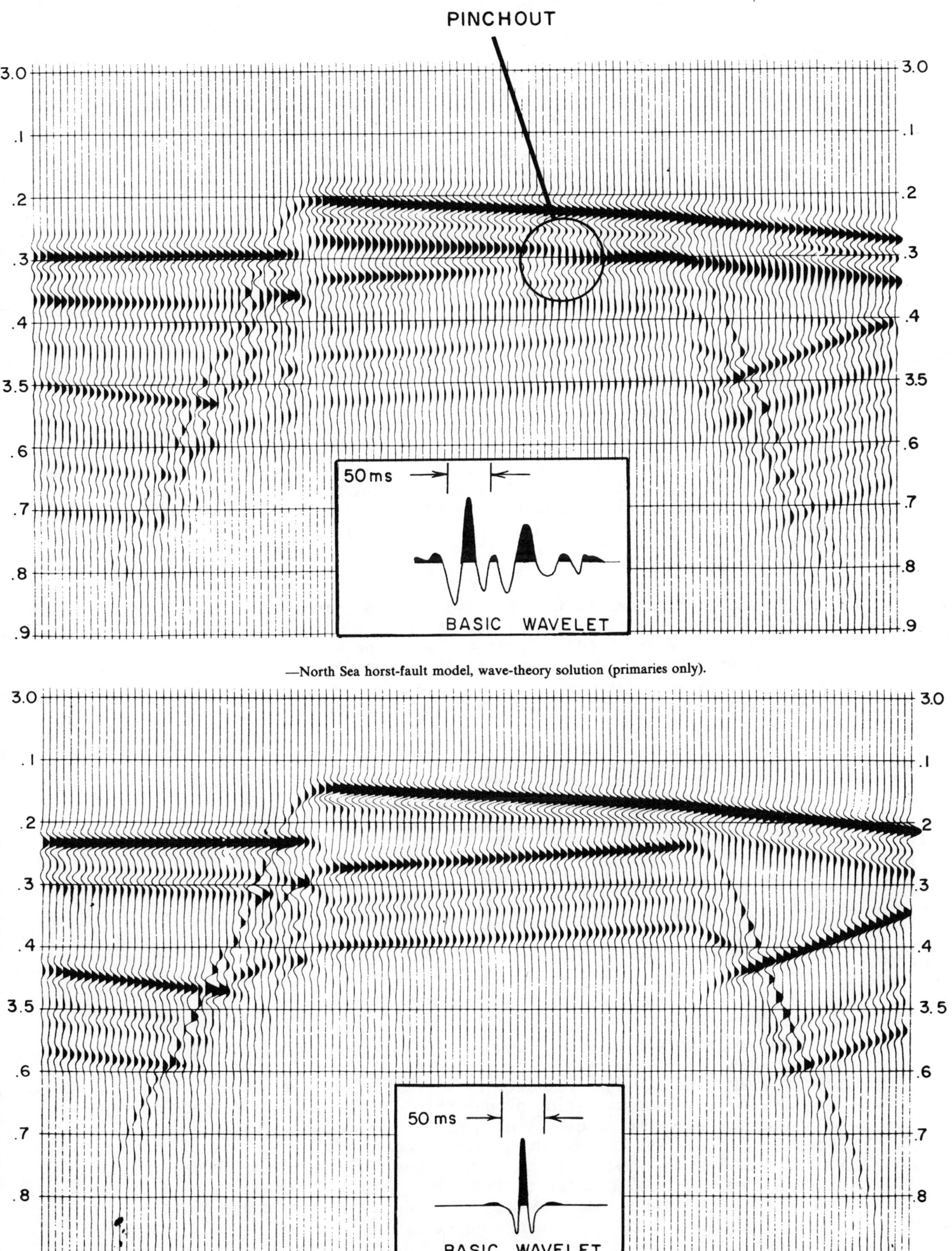

—North Sea horst-fault model, wave-theory solution (primaries only).

—North Sea horst-fault model, wave-theory solution (primaries only).

With the simple waveform, the structural interpretation of the data is rather a simple task. The typical waveform section however, shows a strong indication of a pinchout developing near the center of the horst block. In this case, the superposition of the lengthy waveforms and the slightly discordant geometry have conspired to suggest a stratigraphic element.

The lesson of this illustration is evident and the study goes on to show that despite conventional processing methods, the indication of a stratigraphic propect exists, even when yet another typical but lengthy waveform is introduced. Only the introduction of the symmetric shortened wavelet clarifies the seismic expression.

C. Sand Studies

Two further model studies of specific types of sand bodies are next considered. Earlier, the seismic characteristics of a barrier bar sand were described in the context of illustrating the uses of seismic modeling. These studies by the same authors consider channel sands and shoreline sand lenses.

A geological scenerio is presented for the channel sand and its electric log signature indicated. A transitional top and a well defined base are noted. Turning now to the wave theory seismic response for a typical and an equivalent symmetric wavelet, we note very significant differences in their interpretive potential. In neither case, do diffractions play a role of importance.

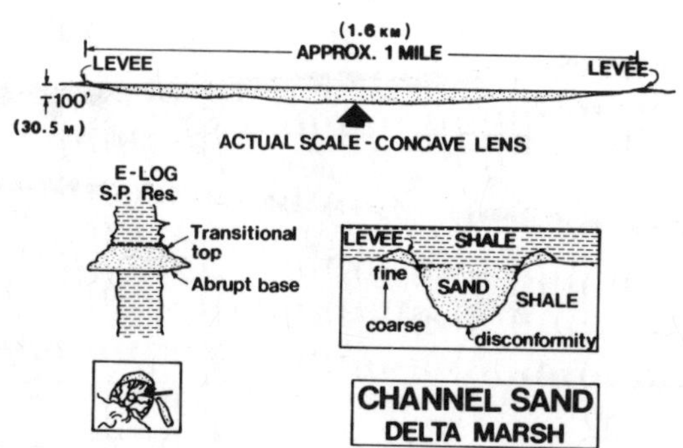

—Geologic model for a channel sandstone.

The relatively thin (100') channel sand has produced a superimposed waveform on the wavelet processed data consisting of a prounounced trough (white) followed by a pronounced peak (black). Here we have the representation of our thin channel sand. Presence of the trough prior to a peak is diagnostic of the low acoustic impedance in contrast to the surrounding shades. The trough is weaker than the peak owing to the transitional nature of the sand top. A time sag is also observed since the principal return is from the base of the channel. We shall defer comment on the estimation of sand thickness until later where quantitative approaches are addressed.

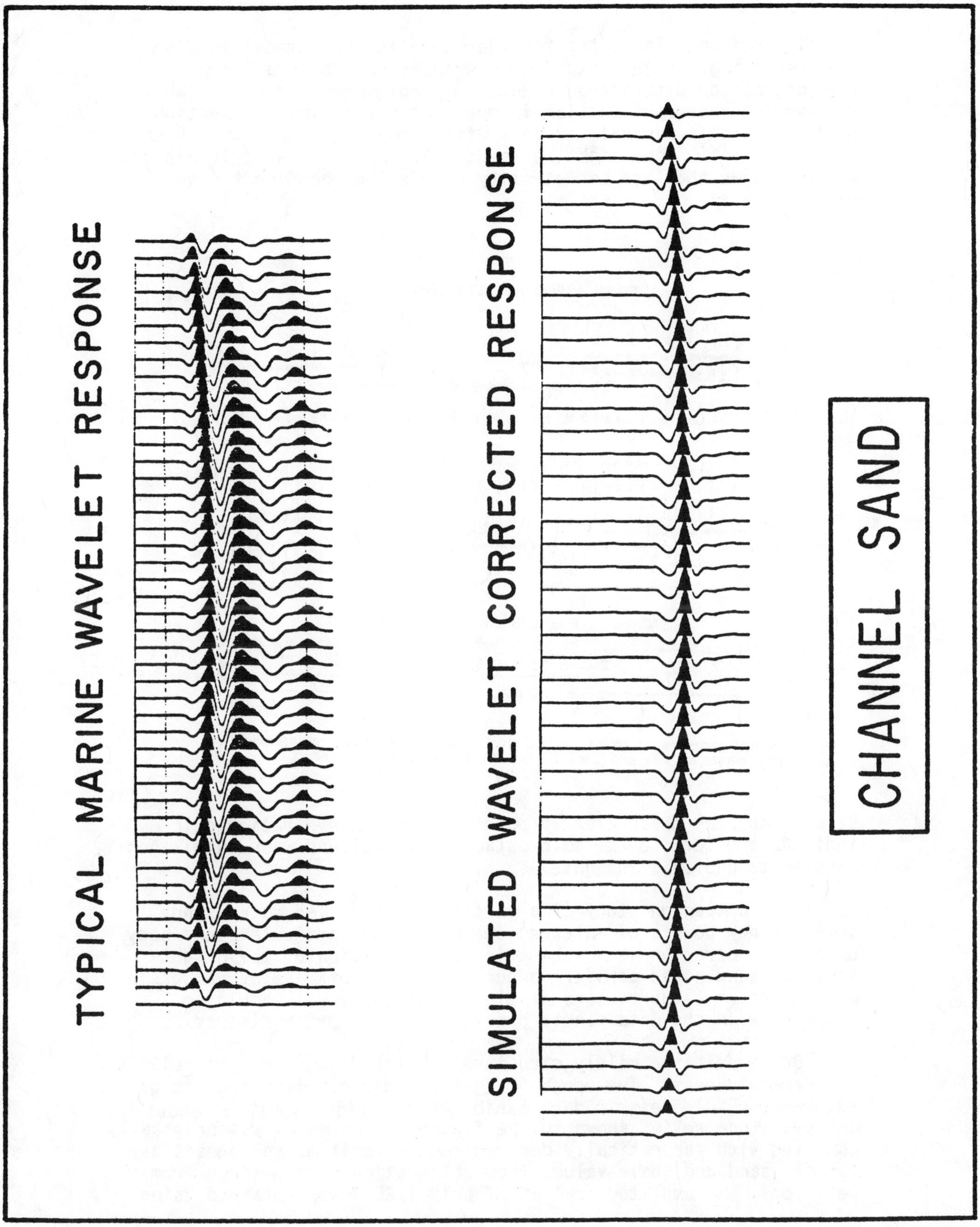

TYPICAL MARINE WAVELET RESPONSE

SIMULATED WAVELET CORRECTED RESPONSE

CHANNEL SAND

The response indicated for the shoreline sand model is also most revealing. In this example no transitional boundaries are present, nor do diffraction events play any important role. While the wavelet processed section is more interpretable, the section developed from a typical waveform offers some complexities. From the usual section one might be tempted to interpret a fault disrupting a sand rather than the permeability barrier between the two en echelon lenses.

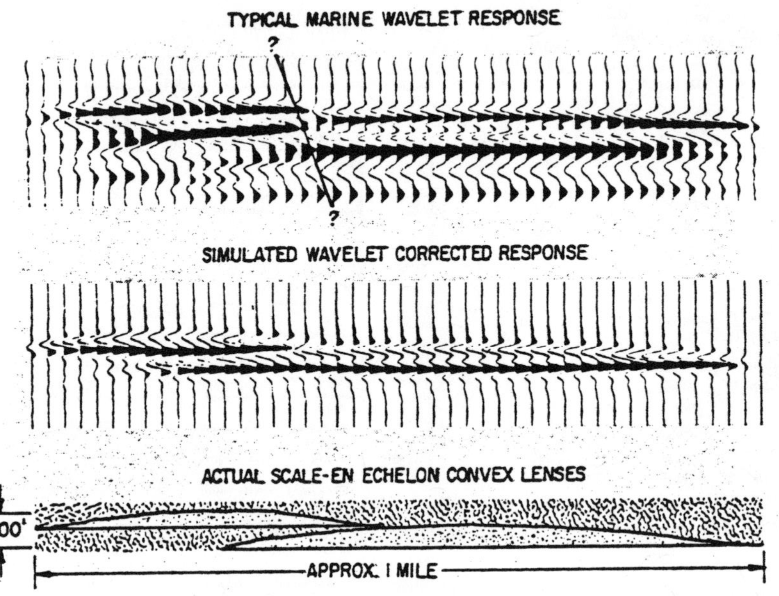

Fig. —Model response of typical marine basic wavelet (A), result of processing for wavelet correction (B) and geologic picture of the sand response (C).

Studies such as these clearly point out that the stratigraphic indicators offered by seismic data can be profound when intelligently coupled to geologic insights.

In a companion study of a partially gas-filled water-sand, some further points of interest emerged. The particular sand body under consideration is a schematic representation of the bright spot gas-sand noted earlier in our discussion of wavelet processing, and the seismic expressions developed for the model should be compared with the data panels presented in the prior discussion.

Our model is a mildly structurally closed sand unit of relative uniform thickness. The upper 60 feet of this 120 feet sand is gas saturated. Field seismic data exhibited a "bright spot" of about the amplitude relief shown in the figure. This model was originally computed with theoretically derived rock velocities and densities for the sand and shale values from calibration data derived from well logs. An amplitude relief of only 1.25:1 was obtained using

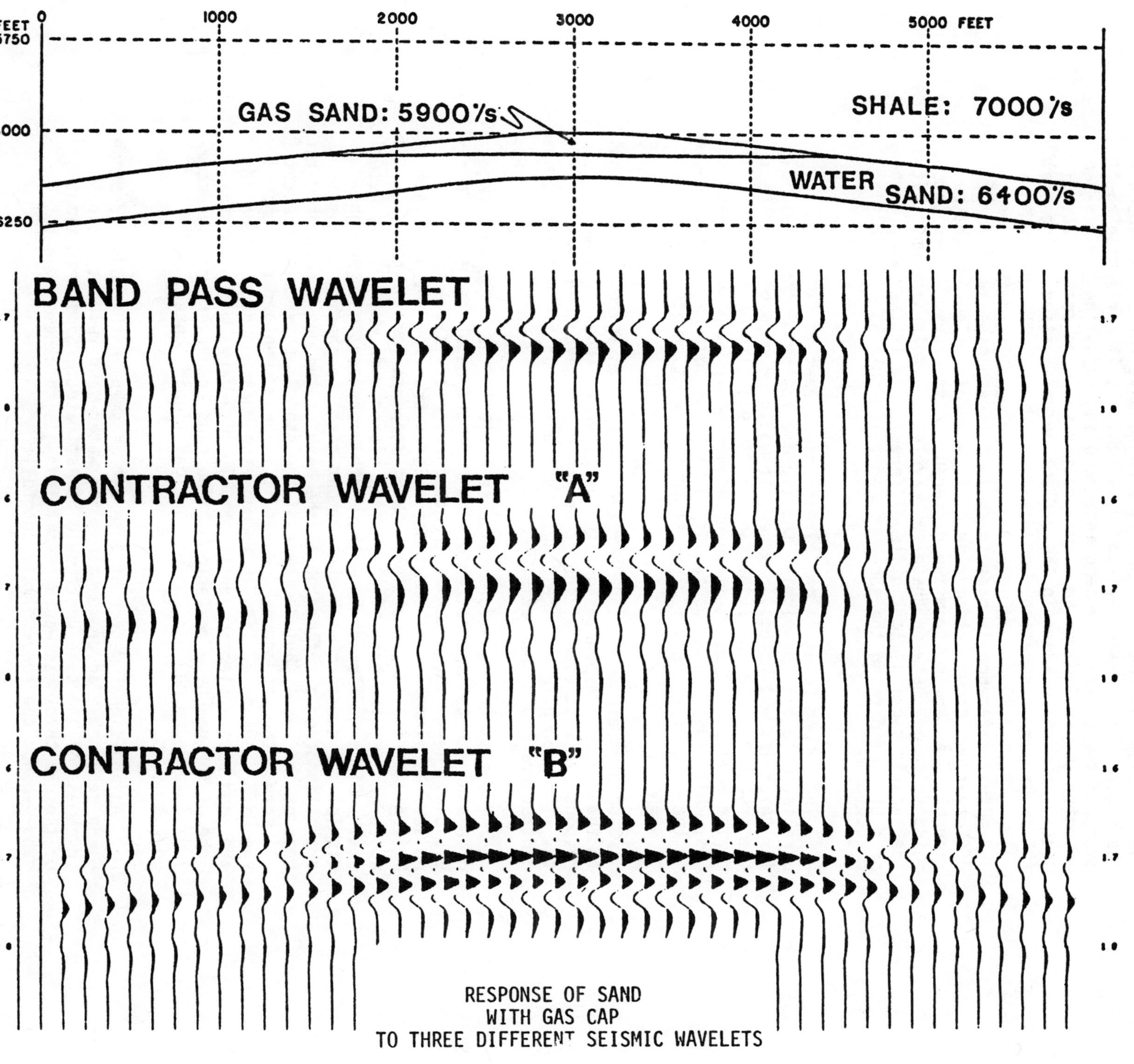

RESPONSE OF SAND
WITH GAS CAP
TO THREE DIFFERENT SEISMIC WAVELETS

these values and the model was altered to have the values shown in the figure. These were derived by assigning a 4% reflection coefficient to the shale-water sand interface and adjusting the gas-sand velocity to provide the amplitude relief shown on seismic data. Densities were included in the reflection calculation and generally follow Gardner's equation.

First reference is made to the model responses shown for two different but documented marine basic wavelets. These might be thought of as coming from two different contractors. It is not our point to choose between these two wavelets. It is apparent that the seismic sections look quite different, and if they were members of a grid of data on the same prospect, it would be difficult to tie them.

The second point is one already made; that it is generally unreliable to attempt picking the top and bottom of the gas-sand for thickness estimates. However, it would be possible to detect the probable presence of the sand and to place it on the map. Marine wavelet "B" might leave us wondering if we have one sand or two, while wavelet "A" causes us to wish for some higher frequencies in the hopes more detail clues as to the exact stratigraphy could be seen.

Both wavelets "A" and "B" have a bandwidth approximately equal to the 8-32 Hz response shown in the figure. Thus, this model response is what each of the other two could have been converted to, with appropriate processing. No problems of tieing data between contractors would then exist.

The symmetric waveform response exhibits the great advantage of having the top and bottom of the gas portion of the sand clearly visible. Next, the water-sand interfaces are visible in this noise-free example, and sometimes this is true for actual data. This gives rise to a question: "What accounts for the phasing appearance at the base of the gas on either side where it is watering out?" Since we already know the depth cross-section, we might dismiss the question with the comment that the phasing is the transition from a reflection at the base of the gas to one from the base of the sand. But the base of the sand is physically present all across the structure, and yet it is only seen out from under the gas. The answer lies in recognizing that the center lobe of the sand base reflection destructively interferes with the side lobe of the gas base reflection. The sand base event simply becomes invisible under the gas because the gas-liquid contact wavelet covers it up.

An additional matter of interest is seen in the model response. Given reasonably accurate values for water-sand velocity, the thickness of the sand unit could be estimated. Similar estimates of the gas thickness would reveal less gas thickness than total sand---hence the reservoir is not full.

Refinements in the nature of the stratigraphic targets which we are able to identify from seismic data require a correspondingly refined use of that data and the available interpretation tools. The studies just described document this viewpoint. In such refined approaches however, information is often forthcoming which bears significantly on still other disciplines. We can demonstrate such a circumstance by considering in more detail some figures which had been treated earlier in another context.

The introductory discussion noted two data panels taken from a North Sea oil field. The upper panel showed a typical processed section which had been subjected to predictive deconvolution before stack. In the lower panel, wavelet processed results had been similarly deconvolved. At the left-hand edge of the wavelet processed result there is an event having geological discontinuity (see arrow) which strongly suggests an oil-water contact. Clearly in this environment, a Jurassic sand under the Kimmeridgian unconformity, an unusual sequence of fluid mechanisms is required to develop sufficient acoustic contrast between the oil and water filled portions so that a seismic event becomes visible.

While the top of the sand is clearly visible, we also see that to the left of trace number 205 some change in lithology is taking place. Beyond the portion of data shown and continuing to the left, the sand top becomes well defined once again with yet another possible indication of an oil-water interface. If the implied lithologic change is indicative of a change in porosity, then serious consequences of a reservoir engineering nature may be further implied.

The ability to define reservoirs with precision and even suggest porosity changes from seismic data are well beyond the span of usually conceived stratigraphic interpretations. We would further note that correlation of the synthetic seismogram with seismic data for the North Sea first shown in earlier discussion corresponds also to this oil field. The Jurassic sand is clearly visible as a low velocity-low density zone on the well logs and is correspondingly denoted by a trough and peak sequence in the seismic data.

D. Three Dimensional Considerations

As a final case study, let us consider a three-dimensional wave theory model study having both structural and stratigraphic elements. The structural contours and an indication of the lithology are indicated on a base map showing also the trace locations for our computed seismic data. Our structure consists of a sand body which transitions into a shale short of the crest. The sand is partially gas filled.

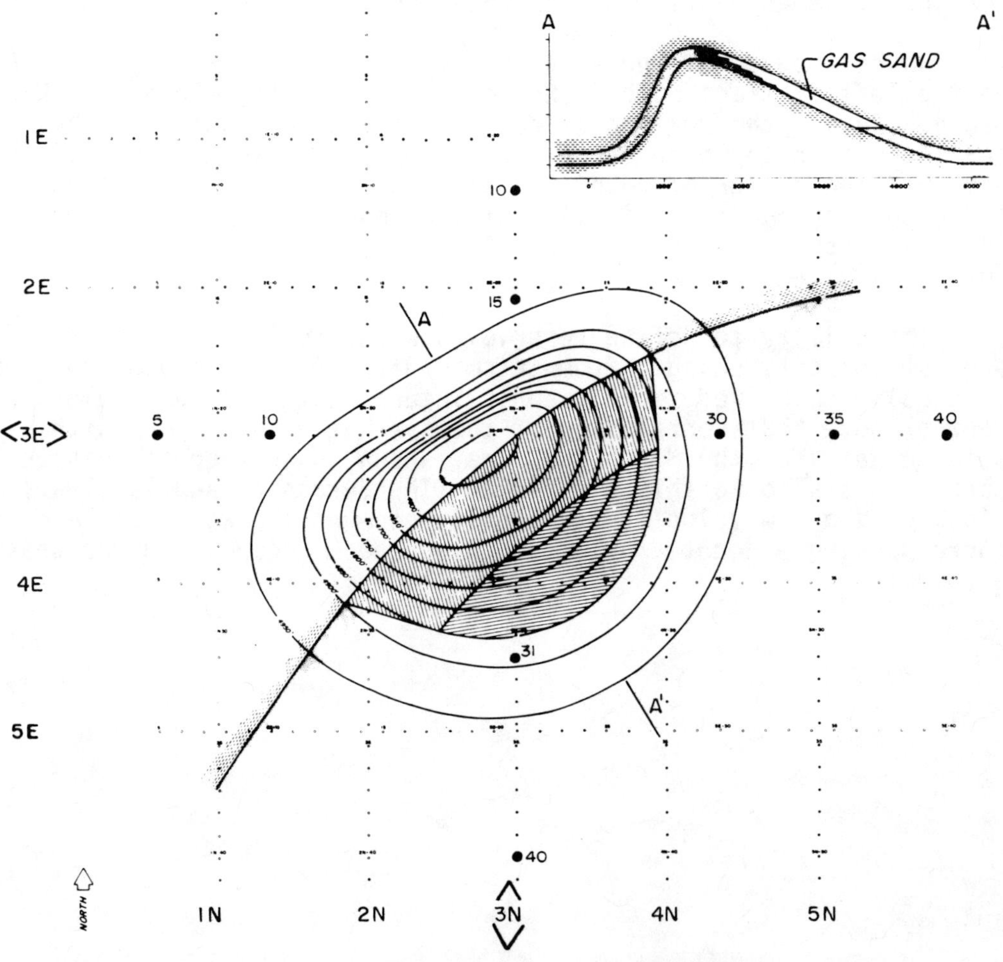

—Three-dimensional model-C depth map.

This model study was conceived and executed by J. P. Lindsey and M. W. Schramm of GeoQuest International, Ltd. using a calculation developed by F. J. Hilterman. The principle lesson of the study is all too obvious. If we honor the structural interpreter's maxim of "drilling on the highs", we will miss the pay zone. Instead we must be guided entirely by stratigraphic precepts entirely.

A first figure shows seismic profiles North-South and East-West across the structure. Since the principal axes of the feature are rotated by about 45° from these directions, our picture is far from straightforward. Still, the most pronounced features of the prospect are the amplitude anomalies to one side of the crest. The principle axes profiles are also shown. While these help considerably in clarifying the interpretation they are usually not available in preliminary exploration studies.

In our discussion of models, we encountered the case of pinnacle reefs in Michigan where structural considerations were paramount despite the stratigraphic elements. Here we have seen a circumstance where the stratigraphic considerations prevail despite the presence of favorable structure.

Catalogues of case studies can only suggest the full potential for stratigraphic interpretation which our tools and techniques place before us. Only our ingenuity and understanding of geological and geophysical principles now limit further developments based on the technology which has been exposed.

While all our previous discussions have been qualitative in nature we shall now want to adopt a quantitative viewpoint in making our stratigraphic correlations. We shall further treat such fundamental questions as the resolution potential of our seismic data in the travel time or equivalently depth sense.

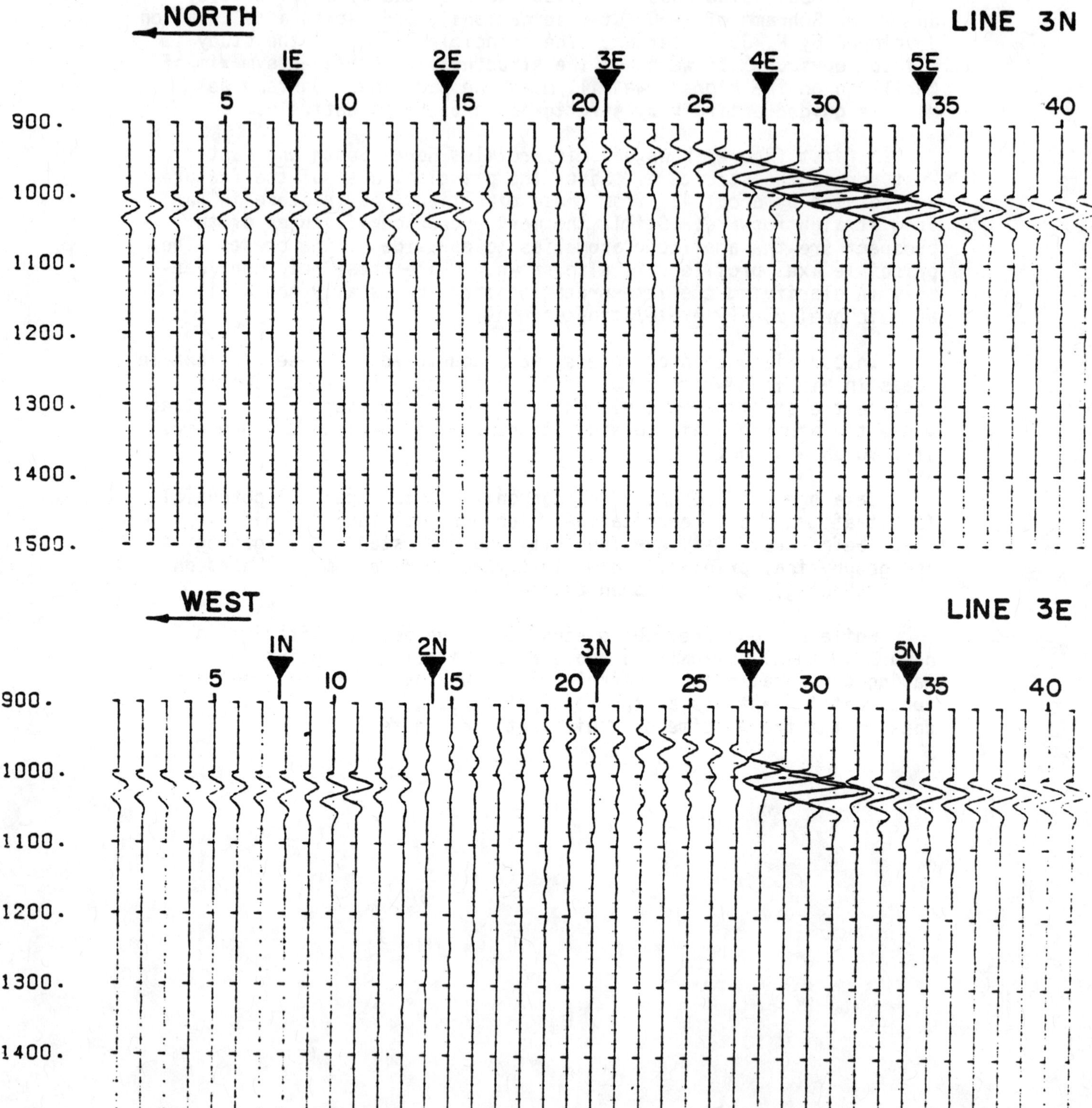
NORTH
LINE 3N
1E
2E
3E
4E
5E
5
10
15
20
25
30
35
40
900.
1000.
1100.
1200.
1300.
1400.
1500.
WEST
LINE 3E
1N
2N
3N
4N
5N
5
10
15
20
25
30
35
40
900.
1000.
1100.
1200.
1300.
1400.
1500.

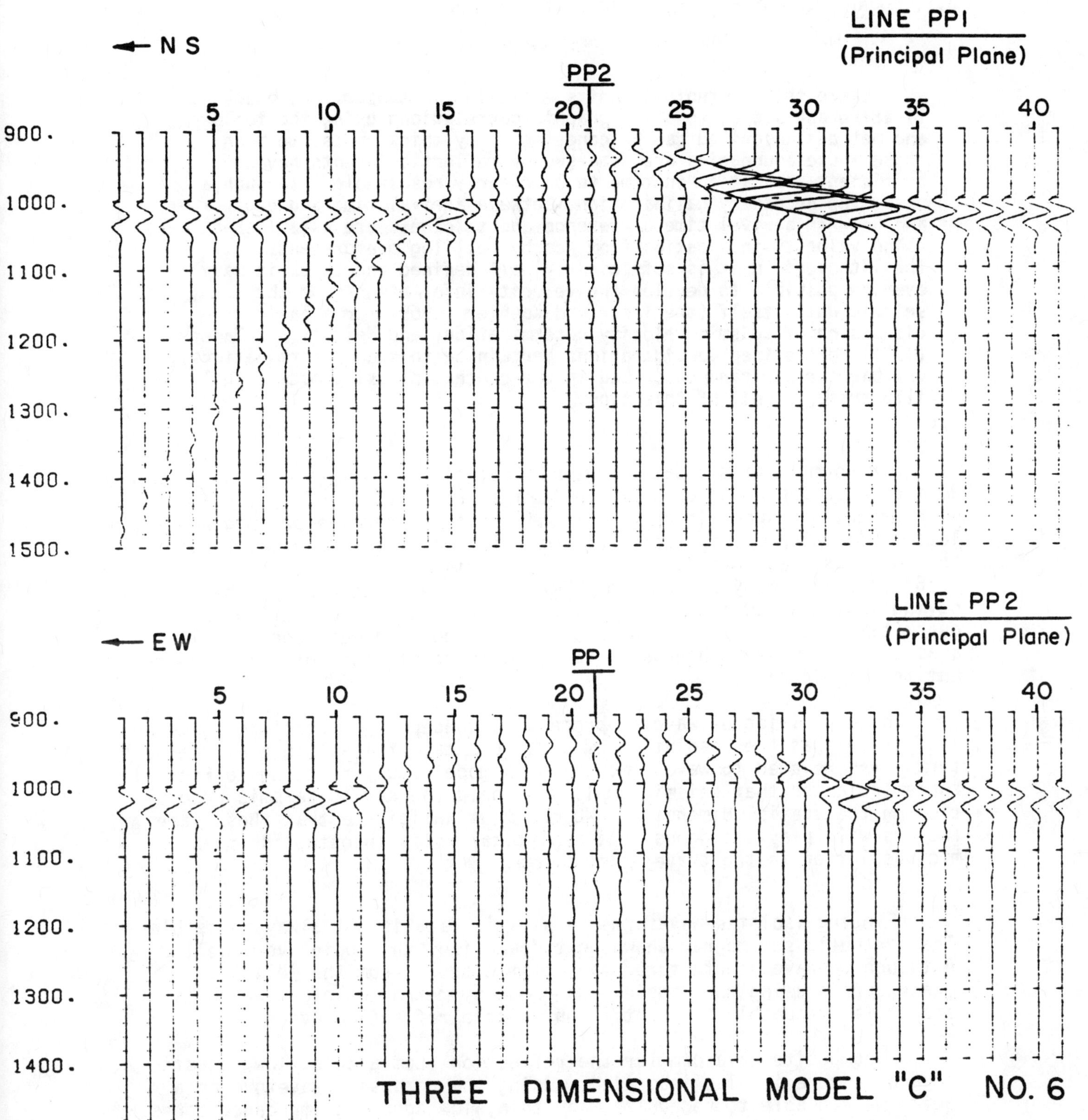

THREE DIMENSIONAL MODEL "C" NO. 6

VII. Quantitative Stratigraphic Correlations

A. Thin Beds and Seismic Amplitudes

Where thick lithologic units exist in the subsurface, quantitative approaches to stratigraphic correlations using the tools and methods described can be conceived. By thick units, we mean those whose boundaries are demarked by reflection events having sufficient separation in time to be clearly resolvable. In such a case, the necessary calibarations between arrival time and depth and between arrival time difference and thickness are established using velocity information from nearby well log measurements. If the lithologic unit is sufficiently well defined and thick it may even be possible to develop the velocity information from the seismic data itself (see Taner and Koehler (1969) for a basic discussion of seismic velocity determination, and Neidell and Taner (1971) for refined considerations pertaining to such determinations). The barrier bar sand described in the context of using models aptly illustrates a unit of this type.

For normal seismic data, thick lithologic units typically encompass 20 m or more depending on depth of burial, regional velocity variation with depth, and specific characteristic of the effective seismic wavelet. This value is, in fact, very much in line with our usual view of the thickness resolution inherent in our seismic data (see Sheriff, 1976). Because many exploration situations are concerned with beds having less than 20-m thickness, the matter of their resolution in quantitative terms is far from academic. Hence, we must consider in analytical terms the thin-bed stratigraphic resolution potential inherent in seismic data.

The exposition of wavelet-processing concepts and practices gave an indication that the resolution of seismic data, at least in qualitative terms, was related to waveform. Also, a model study relating to Fresnel zones indicated that seismic amplitudes held the key to seismic resolution in the spatial dimension. We may thus anticipate that these same factors will play analogous, but analytical roles in establishing seismic resolution in the travel-time sense.

Finding isolated waveforms in seismic data is not always possible. The synthetic seismogram shown next dramatizes this point very well. Although we have manufactured the seismic trace from the well log measurements using the simple zero-phase symmetric waveform which is shown, we are unable to identify any single reflection event.

A composite waveform for the reflection from a thin, low acoustic impedance zone is also indicated. Using the composite waveform as a guide we are able to recognize many thin, low acoustic impedance zones. In the particular geological province of the study these zones are sands sandwiched between higher acoustic impedance shales.

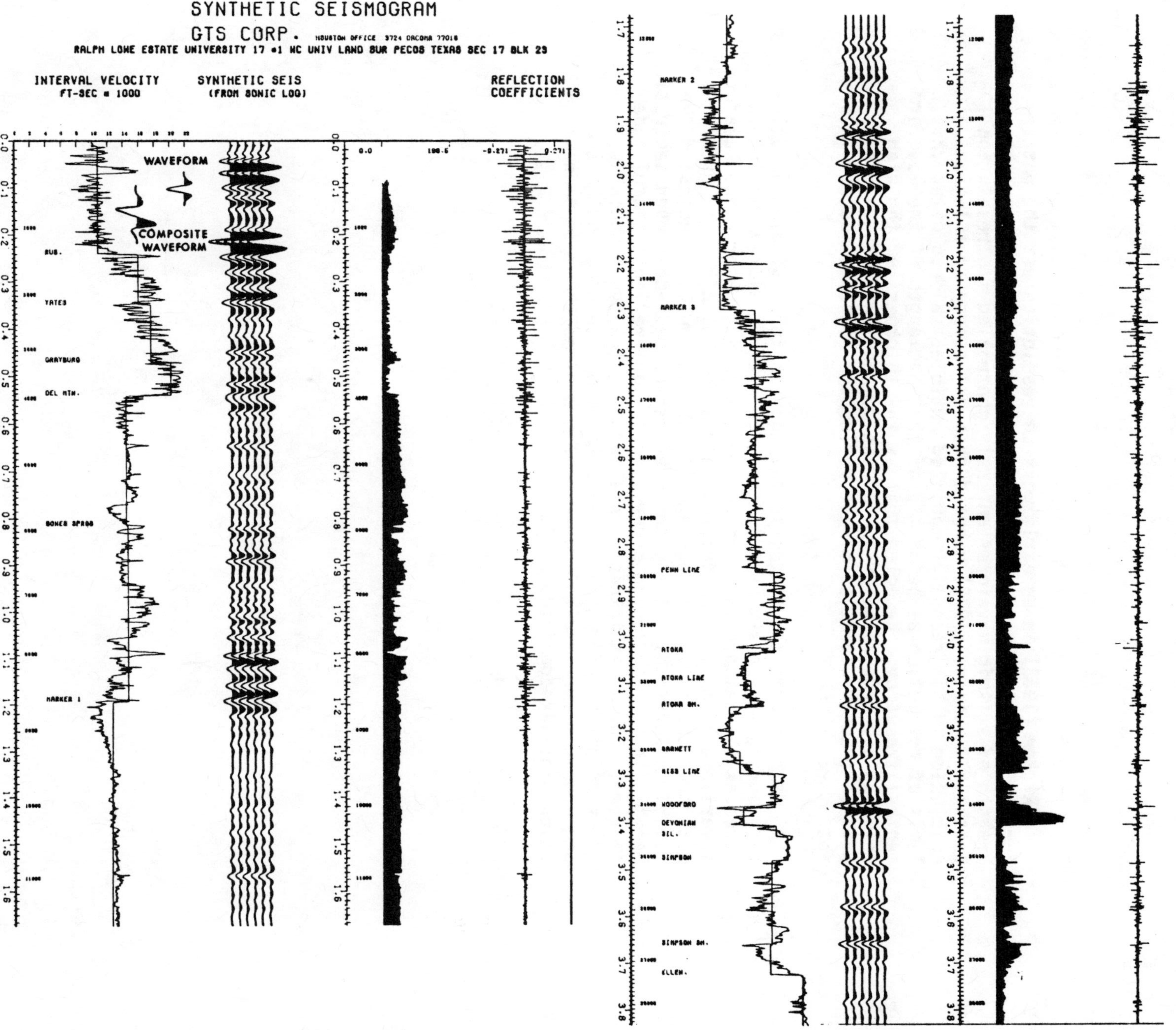

SYNTHETIC SEISMOGRAM
GTS CORP.
HOUSTON OFFICE 3724 ORCOMA 77018
RALPH LOWE ESTATE UNIVERSITY 17 #1 WC UNIV LAND SUR PECOS TEXAS SEC 17 BLK 23
INTERVAL VELOCITY
FT-SEC = 1000
SYNTHETIC SEIS
(FROM SONIC LOG)
REFLECTION
COEFFICIENTS
WAVEFORM
COMPOSITE
WAVEFORM
SUB.
YATES
GRAYBURG
DEL MTN.
BONE SPRGS
MARKER 1
MARKER 2
MARKER 3
PENN LIME
ATOKA
ATOKA LIME
ATOKA SH.
BARNETT
MISS LIME
WOODFORD
DEVONIAN
SIL.
SIMPSON
SIMPSON SH.
ELLEN.

Ricker (1953) in an early work on resolution set the pattern for most of our subsequent thinking. According to Ricker, the breadth of seismic waveforms entirely controlled its resolution potential. Widess (1973) carried this philosophy to a logical conclusion in his recently reprinted article. Widess cited this limit of resolution as being 1/8 of the wavelength of the waveform central frequency. He noted that as the thickness of the high velocity bed dropped below 1/8 of the dominant wavelength, the peak-to-trough separation of the reflection signature became invariant.

We have computed synthetic seismograms for a single low velocity bed for a variety of thicknesses. A 25 Hz Ricker wavelet is taken as the seismic waveform and we do in fact observe the waveform invariance below 40 foot thickness in accord with results of Widess. Note that the peak-to-trough amplitudes are now also indicated. We clearly observe that at the thickness where peak-to-trough separation becomes invariant, tuning occurs and maximum reflection amplitude is seen. For bed thicknesses smaller than the tuning thickness, all resolution information appears to become encoded in the event amplitudes. The results from Widess exhibited the same phenomenon which went unrecognized.

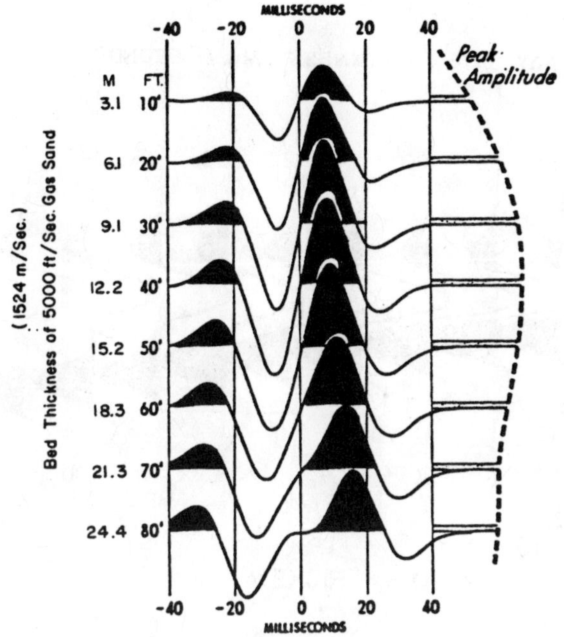

–Response of low-velocity layer—synthetic-seismogram study for 25-Hz Ricker wavelet.

J. P. Lindsey (Geophysical Society of Houston, 1973) explicitly appreciated and utilized the principle of amplitude encoding of thin beds. Of course, in our ealier discussion of the Fresnel zones, we encountered a similar phenomenon in that the amplitudes of seismic signatures for point reflectors related to the spatial extent of small subsurface reflectors.

Hence, wavelet-processed data, taken in conjunction with data in which reflection strengths are controlled so as to have significance, offer a powerful stratigraphic approach to the resolution of thin beds. The practical bounds of the method and specific mechanisms remain to be explained. Still, we must not overlook the profound economic impact on exploration for stratigraphic objectives of this new viewpoint toward seismic resolution.

It is important to stress again that our new viewpoint toward resolution rests in good measure on our knowledge of the propagating waveform and the simplicity of its structure. Where we have no such reference, our insights have far less effect, as is illustrated by the work of Meissner and Meixner (1969). Although thin-bed effects were studied, intuitions received little guidance owing to use of a rather realistic, complex, lengthy seismic source pulse. Clearly, then, we shall want to apply this quantitative technology only to wavelet-processed results in which our requirements for a waveform reference are satisfied.

In the next figure some synthetic-seismogram studies for thin beds are repeated, this time using two differing symmetric waveforms - a 20-Hz Ricker wavelet and an 8 to 32-Hz Butterworth bandpass wavelet. The thicknesses are now presented in two-way travel time units. For every such example a characteristic amplitude tuning thickness is noted, whereas, for lesser thicknesses, all thickness information is amplitude encoded.

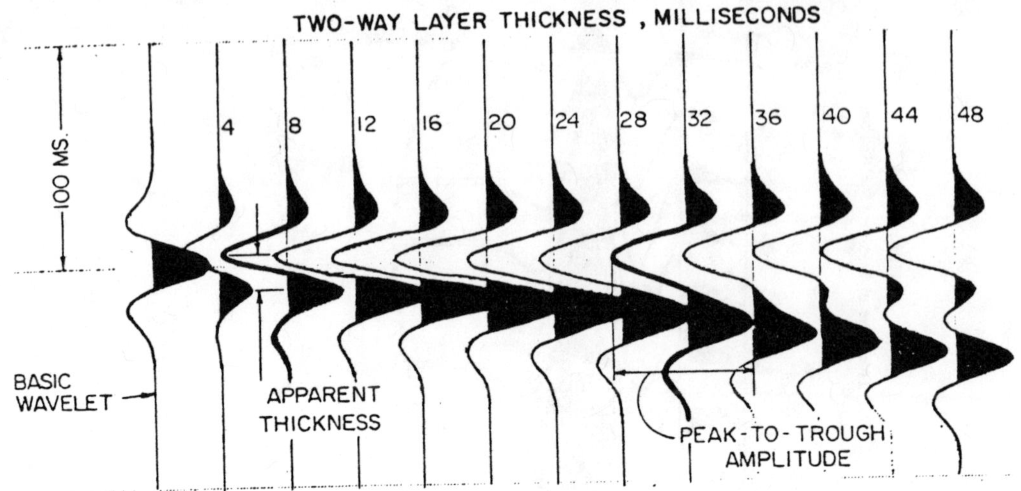

20 Hz RICKER

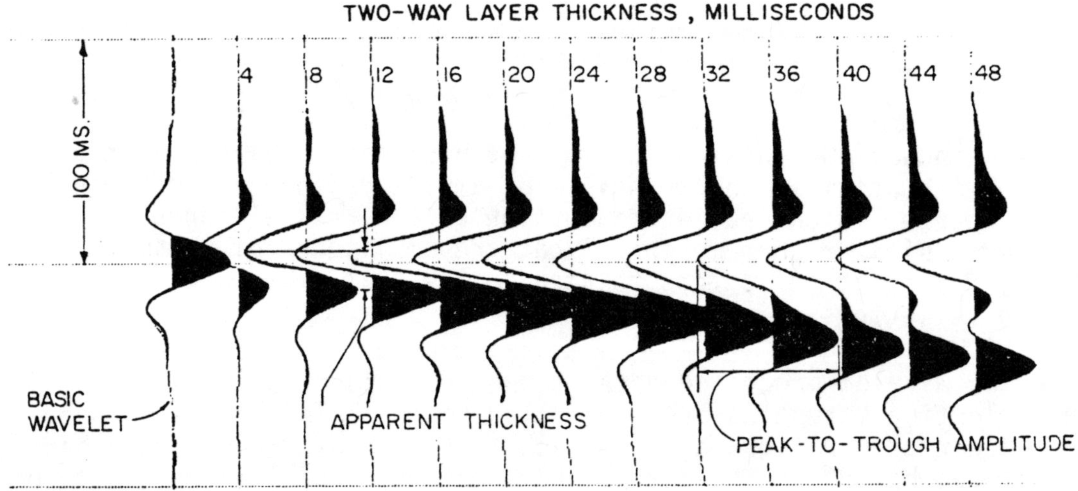

8-32 Hz. BANDPASS
48/48 DB./OCT.
CUTOFF RATE

—Synthetic-seismogram studies of thin beds.

From synthetic-seismogram studies of this type, we can develop sets of calibration curves as are now illustrated. The vertical scale presents actual bed thickness in milliseconds, and the two horizontal scales, in turn, give apparent thickness measured or peak-to-trough time separation and the measured peak-to-trough amplitude in arbitrary scale units. The specific curves shown are taken directly from the synthetic-seismogram study using the 20-Hz Ricker wavelet. If the actual thickness of the bed and the apparent thickness, as determined by peak-to-trough separation, were the same value, the time-resolution calibration curve would follow the diagonal dashed line. For the thicker lithologic units or beds, this dashed line is in fact followed, showing that the resolution in travel time can be accurately accomplished.

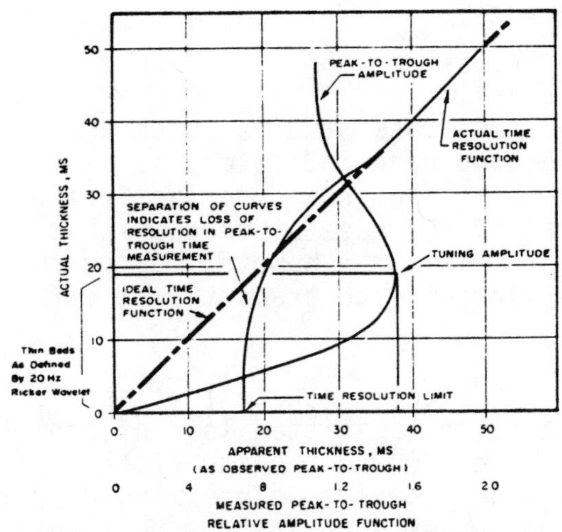

—Resolution calibration curves for 20-Hz Ricker wavelet developed from synthetic-seismogram study.

For the particular 20-Hz Ricker wavelet under consideration, the smallest possible peak-to-trough separation which can be observed approaches an asymptote of about 17.3 msec. Also, at a bed thickness of 19 msec, the indicated seismic strength or amplitude tunes to a value about 40% greater than the normal amplitude level observed for bed thickness of 45 msec or more. It follows that, for seismic data processed to have a 20-Hz Ricker wavelet as the propagating wavelet, thickness estimates for beds having apparent thickness (from peak-to-trough observations) of less than 25 msec are best determined by using calibrated amplitude values.

Recall that the technology for transforming and manipulating seismic waveforms gives us the ability to impress any waveform we choose into the data so long as it does not include frequency components not present in the originally propagating waveform. Calibration curves for amplitude and travel-time measurements as shown may be developed for any specific symmetric wavelet used for wavelet processing. Further, an analytic means now exists for distinguishing between thick and thin beds and resolving thin beds as a function of a known propagating wavelet.

Several practical applications are apparent. First, it is clear that a thinning bed, or one which is pinching out, will have its clearest seismic expression when the tuning thickness is reached. The interpretive significance of this observation is profound for several reasons. At a most elementary level, we see that a thin gas sandstone will have the highest amplitude reflection not necessarily at its thickest portion or where these is most gas, but, rather, where the tuning thickness is attained. This is usually near the edges of the gas-filled zone. Of more consequence to this discussion is the fact that the observation of tuning amplitude levels on the processed seismic section gives us one of two calibration values needed to relate amplitudes on the seismic section to the arbitrary scale of amplitudes on our calibration-curve plot. A second calibration value is normally determined from observation of the normal amplitude level for a thick lithologic unit. The need for preservation of the significance of seismic amplitudes through processing is clearly indicated.

Before we consider some case studies which suggest the power and utility of the technology we have indicated, it is important to note certain limitations of the method which derive from geologic considerations. In fact, several of the case studies which derive from models present such points. At this time, however, using the synthetic seismogram as a tool we may investigate transitional effects in thick beds and thin beds.

In the Figure which follows, we note that as a transitional lower contact thickens, there is an associated loss of amplitude and high frequency components. Note the lengthening of the waveform with high frequency loss and the trough amplitude becoming predominant over the peak. For a transitional upper boundary the peak would be the larger event.

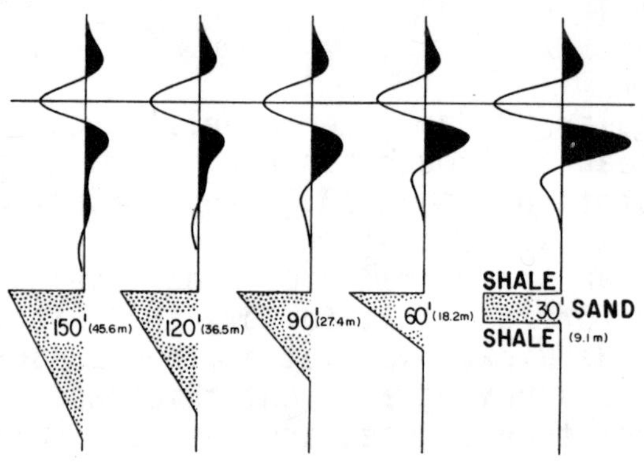

—Seismic response for vertical transition contacts of variable thickness.

Where we deal with a thin boundary, the effect of transitioning is simply to remove acoustic impedance contrast and diminish amplitude. We see only a net loss of material with no clues as to internal distribution.

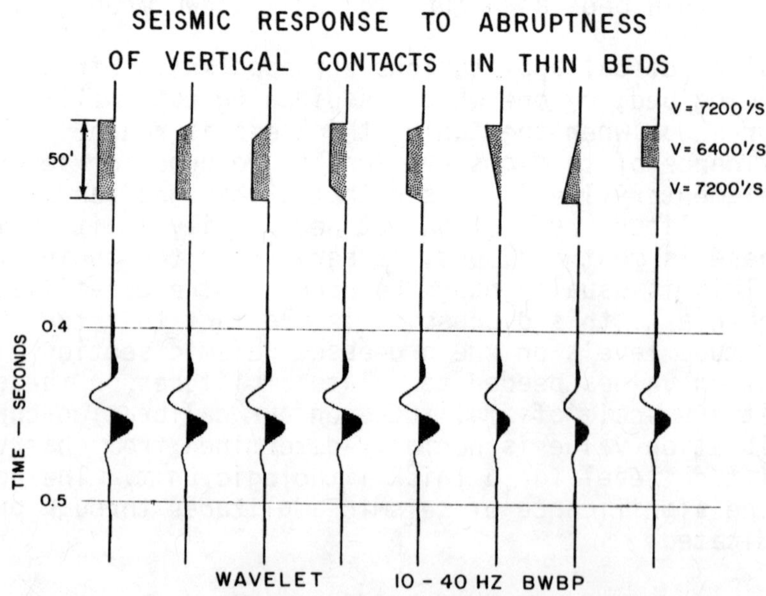

—Seismic response from thin transitional contacts (wavelet 10 to 40-Hz BWBP).

B. Thin Bed Model Case Studies

Next we consider a wave-theory response from a schematic model after Lindsey et al (1976), which shows a thin shale stringer in a sandstone body of uniform thickness. The 15.2 m of the gross body character leads to an invariance of waveform. Net shale content can be estimated directly by noting the peak-to-trough amplitudes.

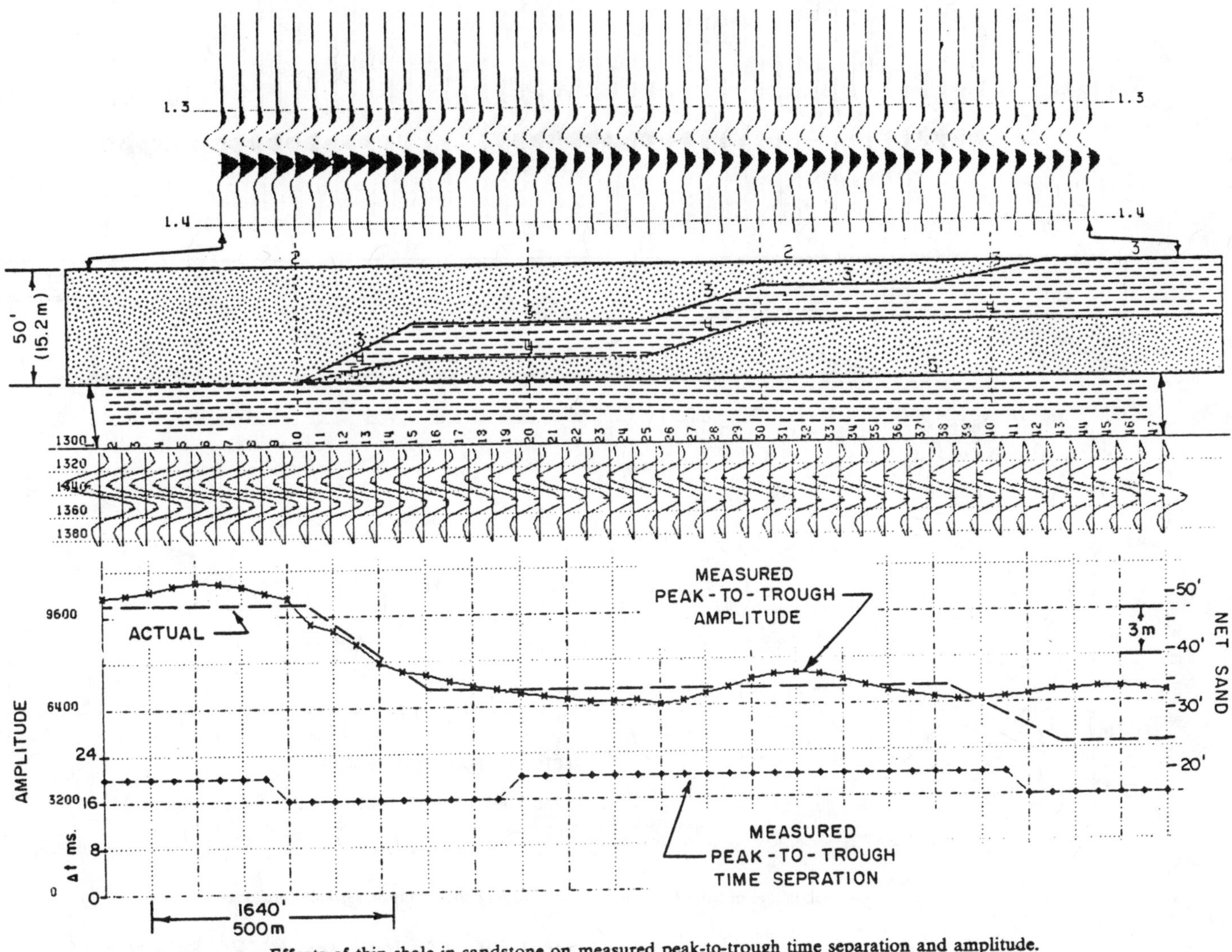

—Effects of thin shale in sandstone on measured peak-to-trough time separation and amplitude.

At the bottom of the figure, an output from a computer analysis monitors the measured peak-to-trough time separation and peak-to-trough amplitude on a trace-by-trace basis. Excellent agreement is noted between the net sandstone estimates and the measured amplitudes. The peak-to-trough time separation in this case tells us nothing because we are dealing with a thin bed. Note that the most serious departures of the measured net sandstone curve from the actual curve correspond to places in the model where beds terminate. These terminations give rise to diffraction events and contaminate the reflections, causing departures from the theoretical considerations which were presented, as the latter were based on the more simplified theory of synthetic seismograms.

A second schematic model study again emphasizes these same considerations. In the following figure a more realistic case is presented in which a transitional sandshale facies is shown. Here we explicitly note the equivalence in seismic expression of two thin beds to one thicker one in terms of amplitude effect. Hence any insights into lithologic distributions in thin units must derive from geologic concepts and principles applicable to the circumstance. They are not directly provided by the seismic data.

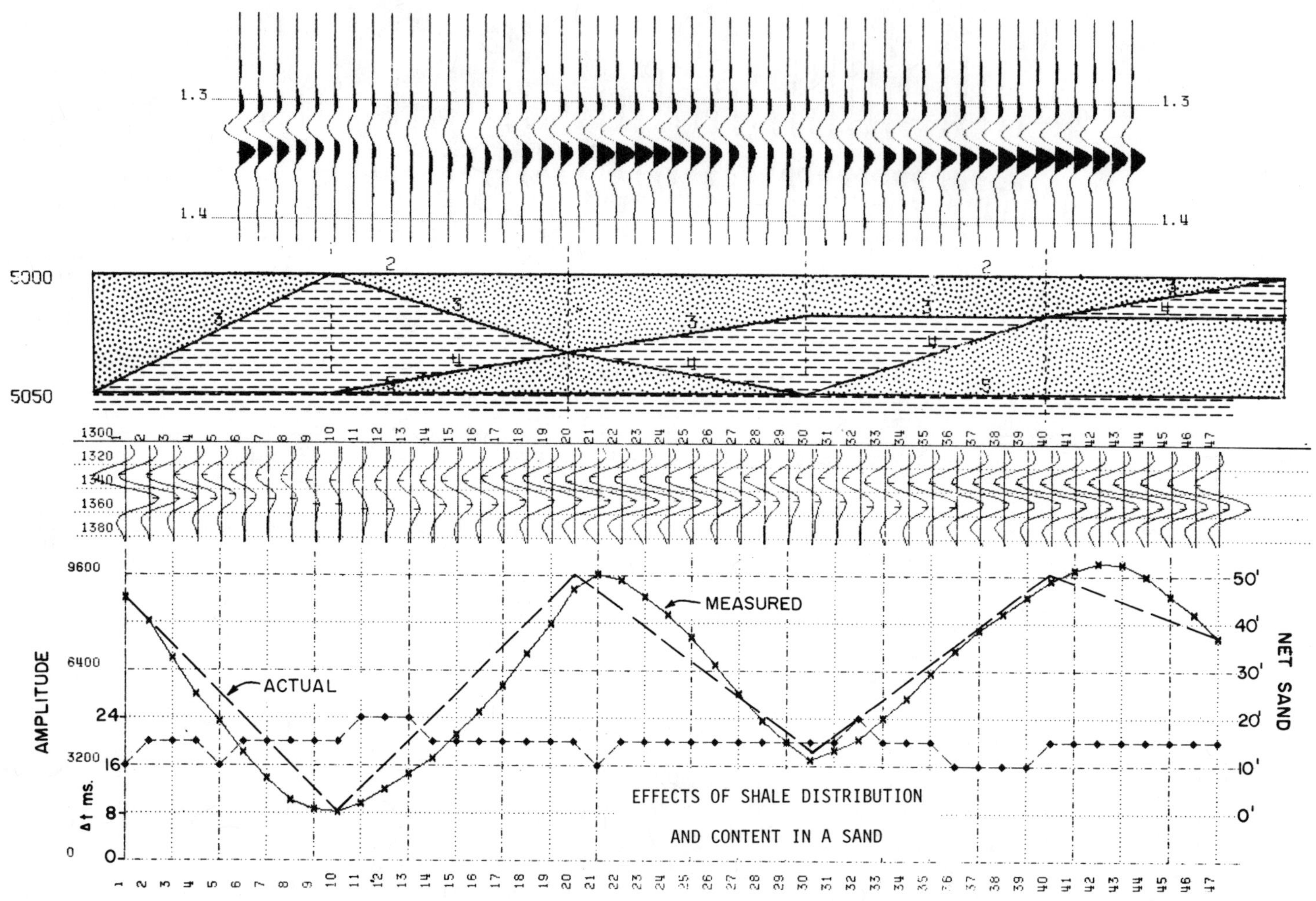

—Seismic model of sandstone reservoir containing shale unit of variable thickness.

We consider next a model by J. P. Lindsey and M. W. Schramm of transitional lithology against a structural background similar in concept to the three-dimensional model study considered in the context of qualitative correlations. A fault has been introduced in place of tight curvature at the left hand edge of the model. Once again we put to practice our computer analysis and determine gross lithology, but now in the presence of a structural component. Note that amplitude level of the water sand has been interpreted and that all amplitude variation exhibited is of lithologic origin and not related to thinning or thickening of the particular unit.

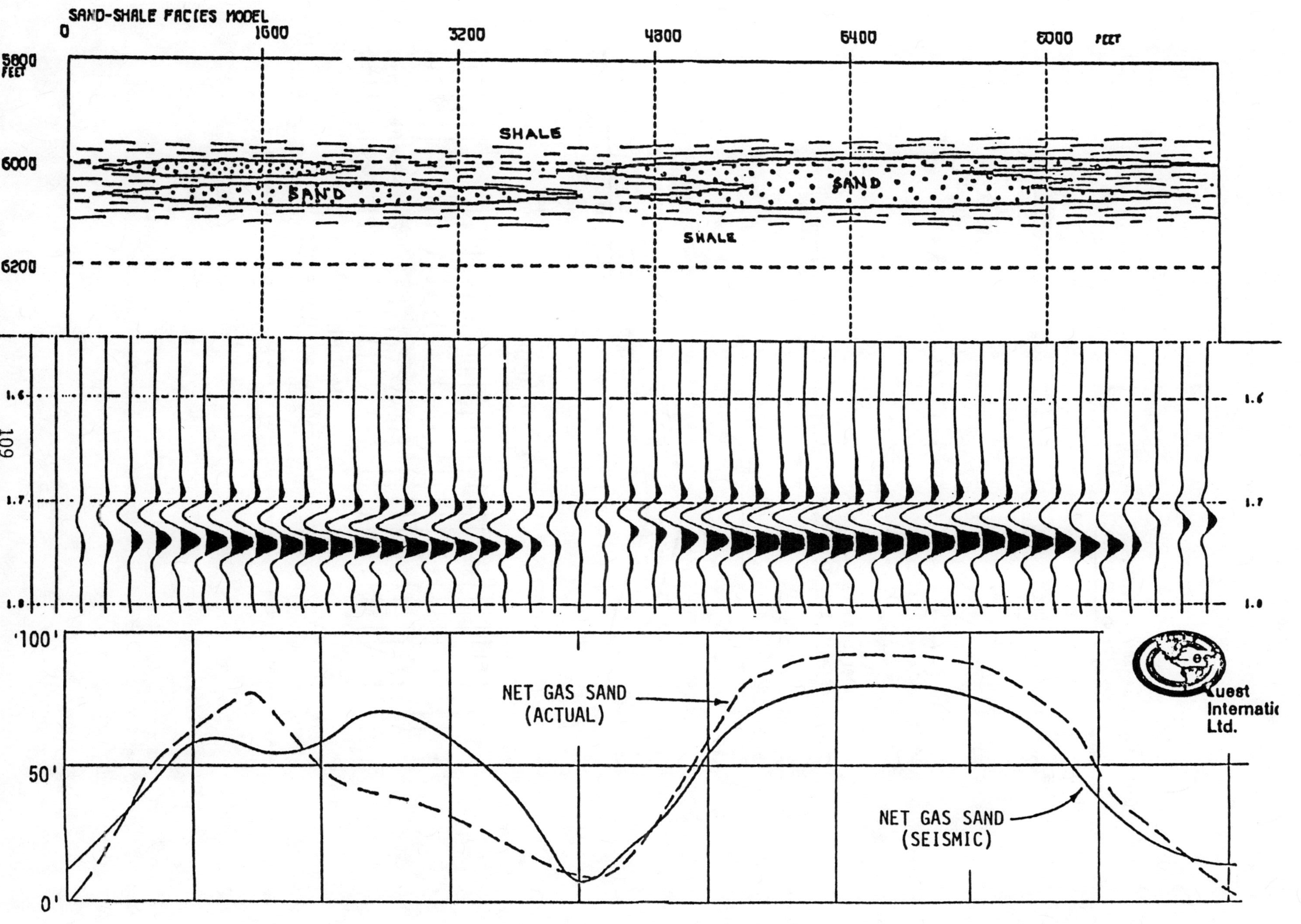

SAND-SHALE FACIES MODEL

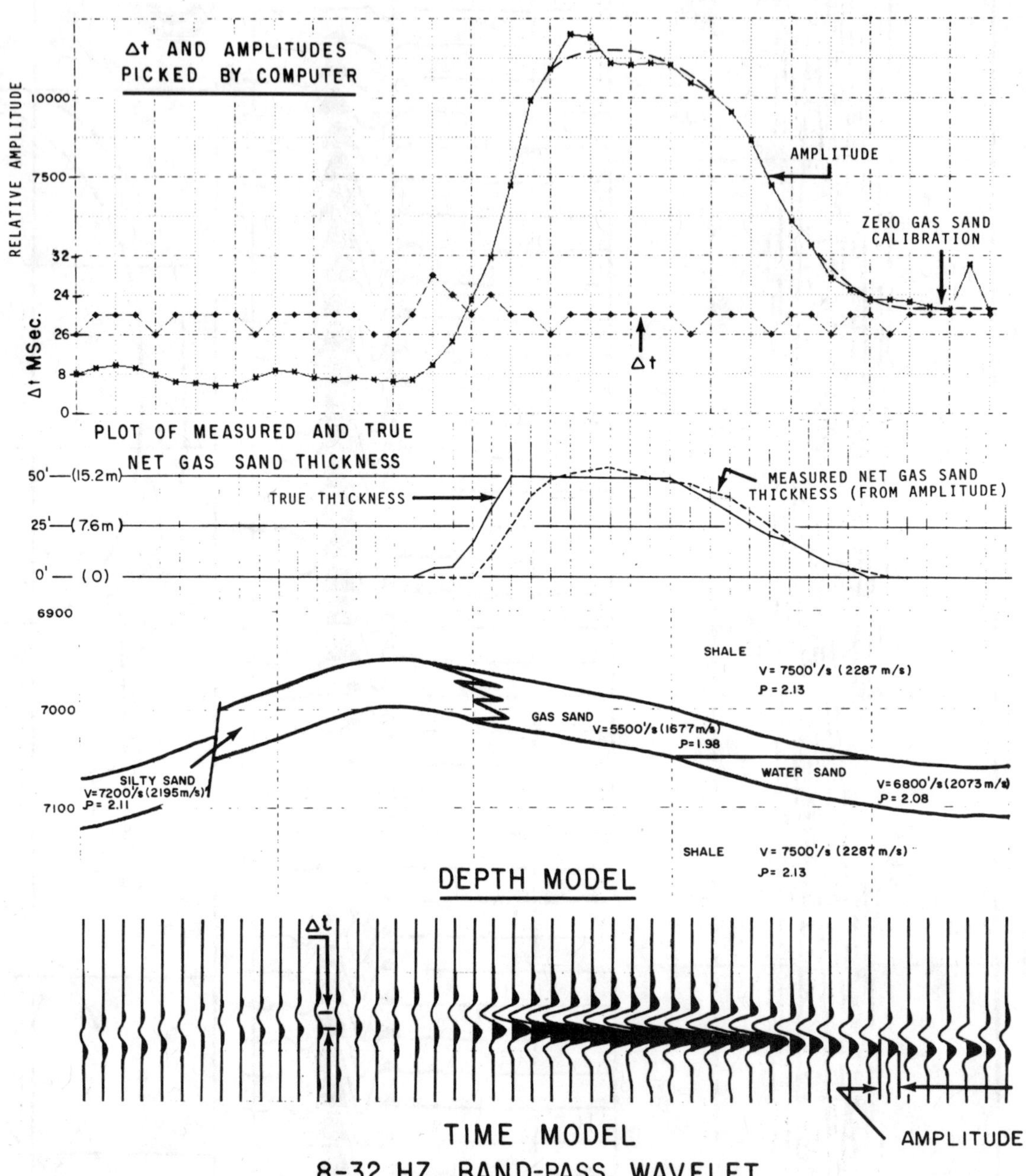

—Updip Permeability Barrier: seismic amplitude as indicator of net hydrocarbon thickness, loss of porosity, and gas/liquid contact.

Before going any further it is essential that several considerations be emphasized once again. First, when we detect thin bed lithology we are looking only at <u>gross</u> lithology and have no information about its distribution or spatial position within the host lithology. The first two model studies illustrated this point very well. Next, the technique we are describing works very well in the case where we have a low (or high) acoustic impedance material inbedded in a uniform host material. Of course in the case of high acoustic impedance the peak event will preceed the trough in reflection time. Other geologic scenarios will require the development of analogous procedures. The techniques being illustrated here for sandstone reservoirs encased in shale can not be universally applied. We shall give further discussion to this matter when we look more closely at procedures.

C. Data Studies

Turning to real data at this time admits essentially the same type of result. In particular we consider the Gulf Coast gas sand of the wavelet processing discussion. In the following illustration, this sand is subjected to the computer analysis of timing and amplitude on a trace by trace basis. Somewhat greater variability is exhibited by the results owing to the noise content of the data. Hence some degree of smoothing is introduced in order to interpret the results.

The invariance of Δt (peak-to-trough) time separations over most of the data analysis output is a first and strong indication that a thin lithologic unit is under study. Peak-to-trough amplitudes on the other hand do appear to tune to a peak value in the vicinity of trace number 140. These values also settle to a plateau level at trace 154 and remain at this level to the right of the data panel. In interpreting these results we must make use of both geologic principles and subsurface determinations which are available to us.

To the right of the data panel we interpret the amplitude plateau as commencing at a gas-water contact and denoting a water-filled portion of the sand body. Using regional values developed from other information a reflection coefficient of 0.044 is assigned as the reflection coefficient between the water sand and the surrounding shales.

For the 8-32 Hz band-pass wavelet introduced by data processing, the tuned amplitude level corresponds to a calibration level of 1.56 on a suite of curves analogous to those shown earlier for the 20 Hz Ricker wavelet. Using regionally applicable velocities for gas sands, we determine that the thickness equivalent to the indicated time thickness at the tuning amplitude is about 54 feet. The reflection coefficient with the surrounding shales in the light of this velocity information and the tuning amplitude factor is estimated to be 0.165 for such thickness. Hence a calibrated scale of reflectivity is developed, as is a graduated lithologic unit thickness scale in terms of peak-to-trough amplitudes. The latter scale however requires use of the calibration curves appertaining to the particular wavelet. A zero amplitude value on that graduated scale should correspond to a zero thickness of gas-sand provided that Δt remains at the constant level and the lithology remains essentally unchanged.

At the right hand of the gas sand we encounter the water contact and transition laterally to the water-filled sand - a change in lithology. With the spatial resolution inherent in the computer analysis we observe a minimum gas-sand thickness of about 15 feet rather than zero. This value arises principally by lateral averaging over a Fresnel zone as was discussed. We can interpret the location of the zero gas-sand thickness quite reliably from both the data panel and the analysis and verify our finding with an

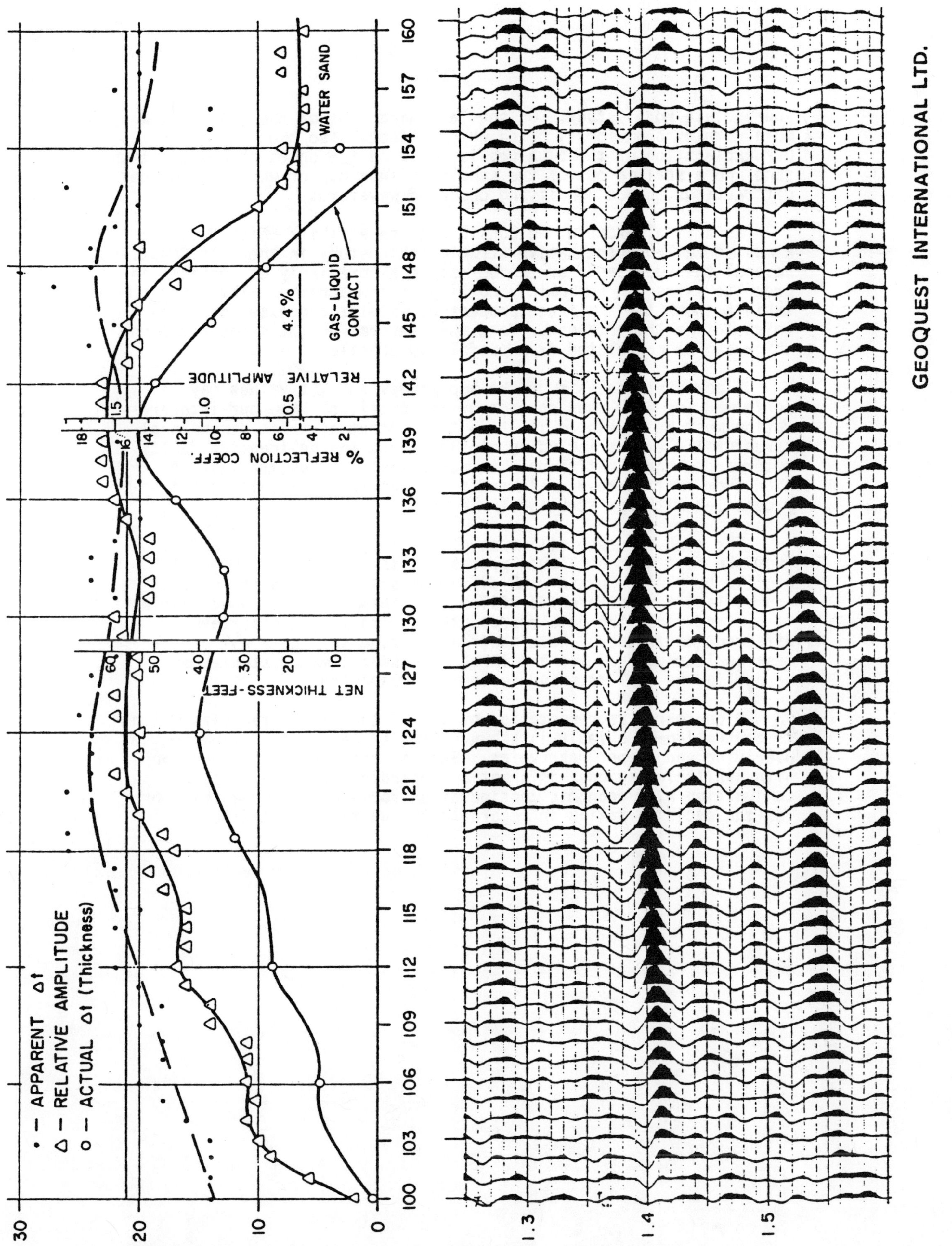

• – APPARENT Δt
Δ – RELATIVE AMPLITUDE
o – ACTUAL Δt (Thickness)

MILLISECONDS

30
20
10
0

NET THICKNESS-FEET

60 50 40 30 20 10

% REFLECTION COEFF.

18 16 14 12 10 8 6 4 2

RELATIVE AMPLITUDE

1.5 1.0 0.5

4.4%

GAS-LIQUID CONTACT

WATER SAND

100 103 106 109 112 115 118 121 124 127 130 133 136 139 142 145 148 151 154 157 160

1.3
1.4
1.5

GEOQUEST INTERNATIONAL LTD.

appropriate model study. On the left hand side of the gas sand, a more complex behavior is noted which affects also the Δt values. Our explanation of these values is based on the onset of a transition zone and a shaling out of the sand body in a manner which introduces diffractions, hence, modifying observed waveforms.

As a final example, we consider three exhibits which describe a most detailed quantitative stratigraphic study in the Gulf Coast. Several thin gas sands are to be mapped as precisely as possible in order to asses the economic viability of the acreage. The first exhibit shows the amplitude processed, wavelet processed profiles which define the prospect. Note that several bright sands are seen which can be correlated from profile to profile.

A computer analysis of the amplitudes and Δt values may be performed for each sand on each of the profiles. Now our interpretation must be developed first in the context of three-dimensional structural elements including closing loops and delineating faults. Afterwards, the stratigraphic computer analyses results must be tied from profile to profile and the quantitative results obtained reconciled with the structural viewpoint.

The need for reconciliation of the computer analyses arises from the fact that amplitude processing is usually performed on a profile by profile basis with different scale factors often being appropriate for different profiles. In this example, added complexity is present since the data represents results from two surveys by different contractors. Note that in treating this data all of the lessons learned from two and three dimensional model studies must be applied as well as sound geological reasoning.

A contoured representation of one of the sand reservoirs which is controlled by a fault is shown. Similar maps will delineate the overlying and deeper reservoirs. Such information in conjunction with available regional information regarding porosity, etc, and in conjunction with economic consideration forms as rational framework for the decision making in regard to the particular acreage.

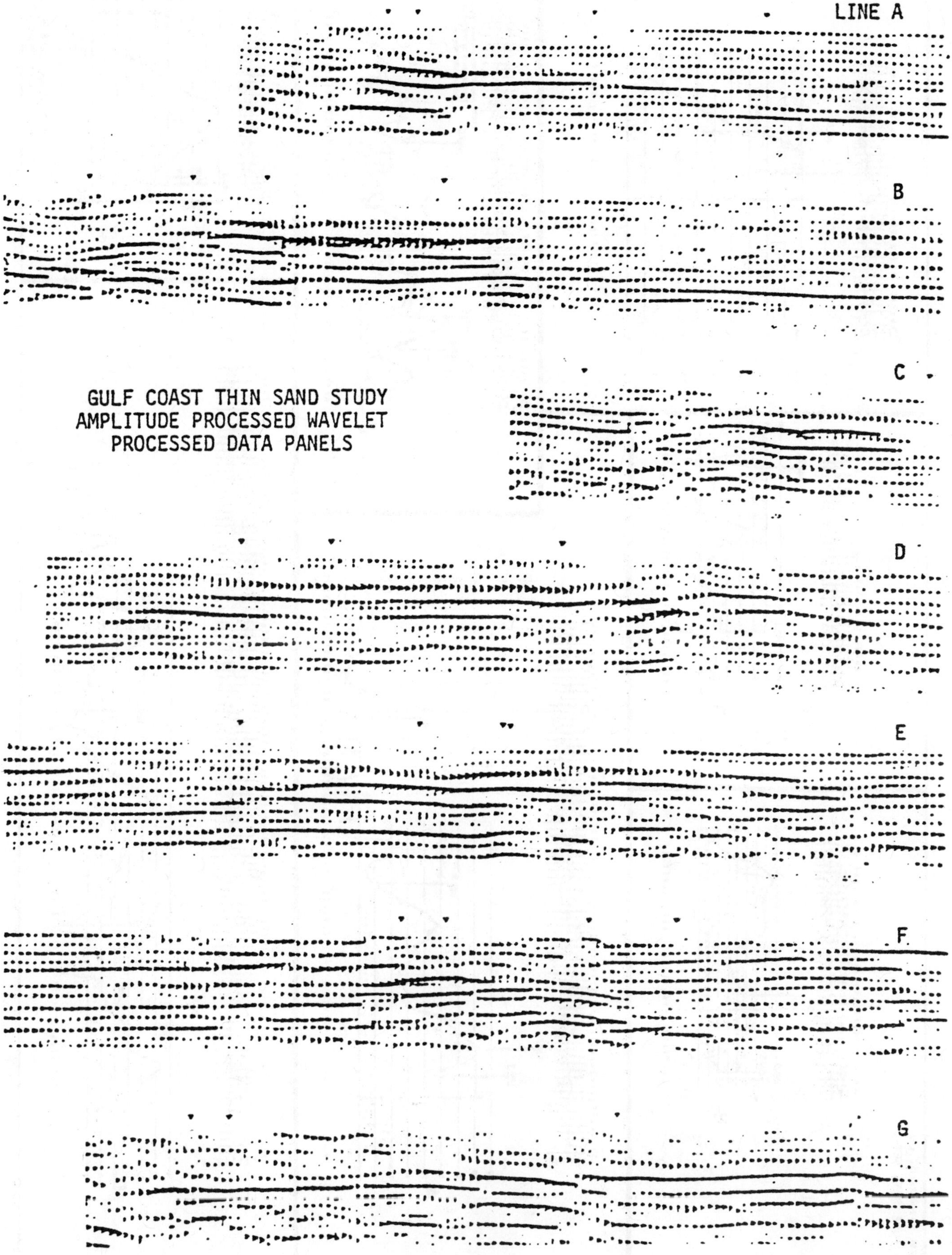

LINE A

B

C

GULF COAST THIN SAND STUDY
AMPLITUDE PROCESSED WAVELET
PROCESSED DATA PANELS

D

E

F

G

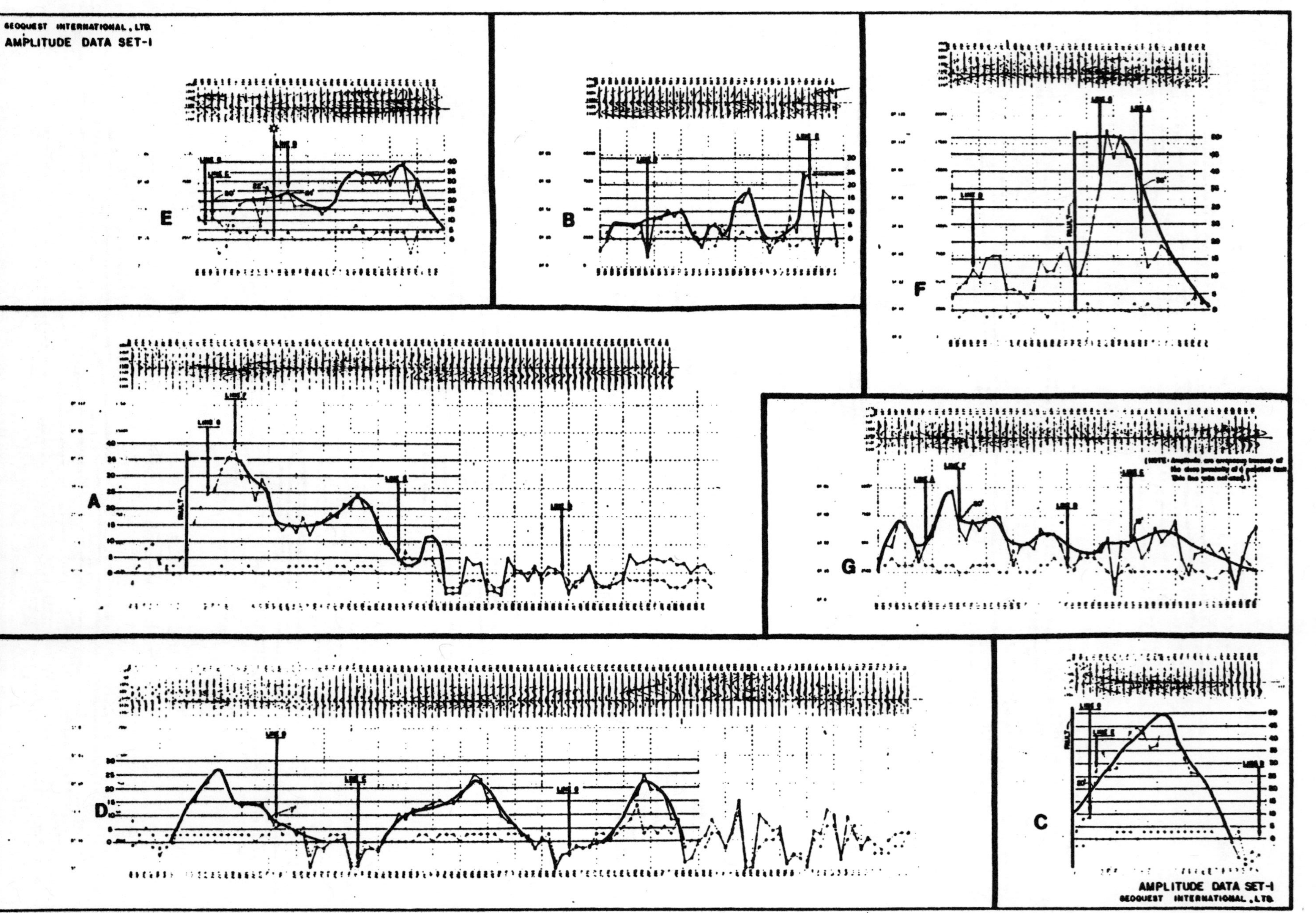
GEOQUEST INTERNATIONAL, LTD.
AMPLITUDE DATA SET-I
E
B
F
A
G
D
C
AMPLITUDE DATA SET-I
GEOQUEST INTERNATIONAL, LTD.

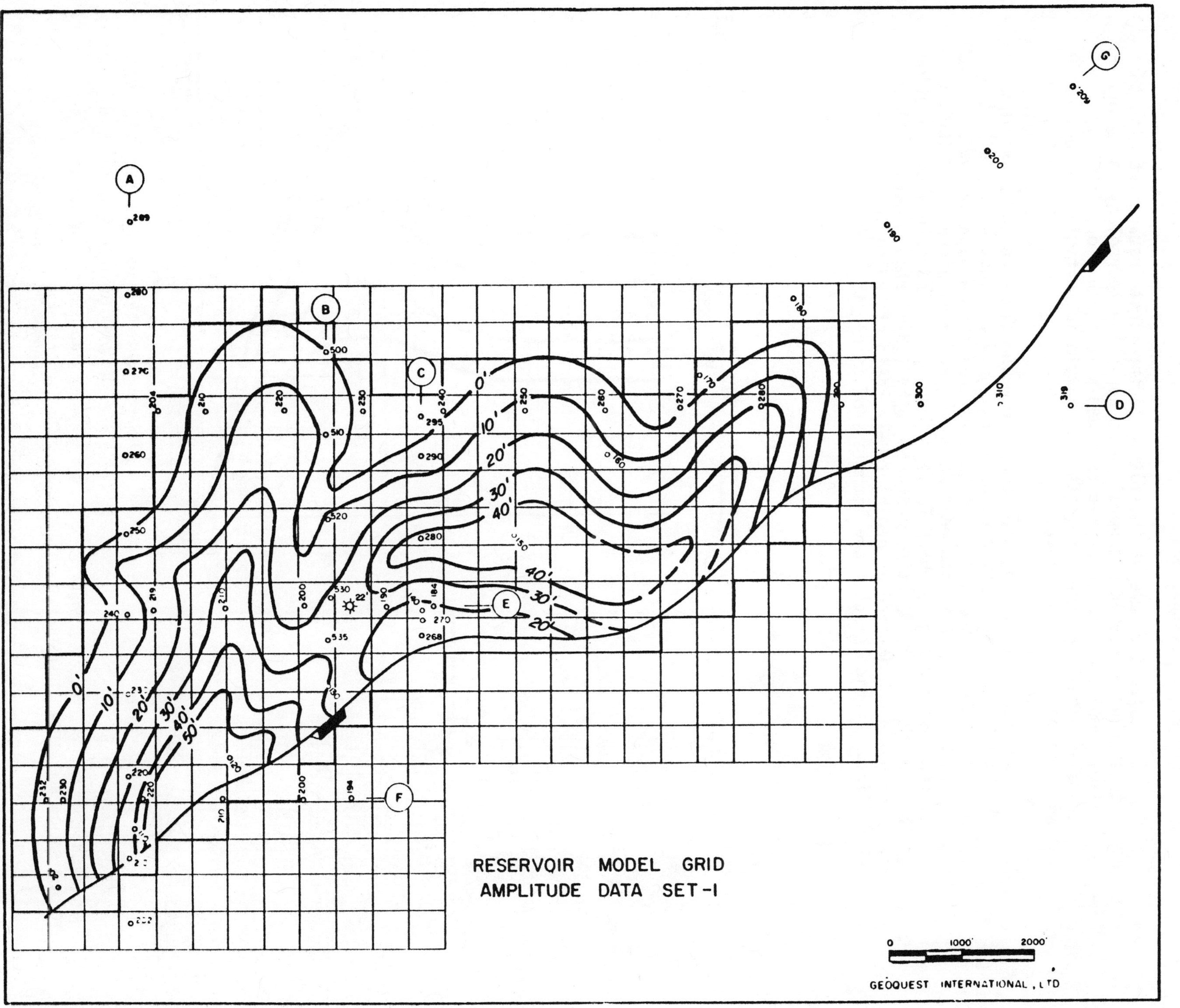
RESERVOIR MODEL GRID
AMPLITUDE DATA SET-I
0
1000'
2000'
GEOQUEST INTERNATIONAL, LTD

D. Procedural Discussions

Our objective here is two-fold. In the first instance we wish to suggest how one may in a general sense apply the technology for using seismic amplitudes to interpret thin bed stratigraphy or lithology. Next, we wish to present at least one example in sufficient detail to serve in an instructive role.

We have already suggested that the methodology which we use for thin sand bodies in a uniform shale environment is not universally applicable. For example, suppose that we deal with a high or low acoustic impedance bed in an environment where the material above and below are different. The Figure which follows from Meckel and Nath's contribution to AAPG Monograph 26, Seismic Stratigraphy (1977) shows that amplitudes must now be calibrated according to the ratio of the reflection coefficients which bracket the lithology of interest.

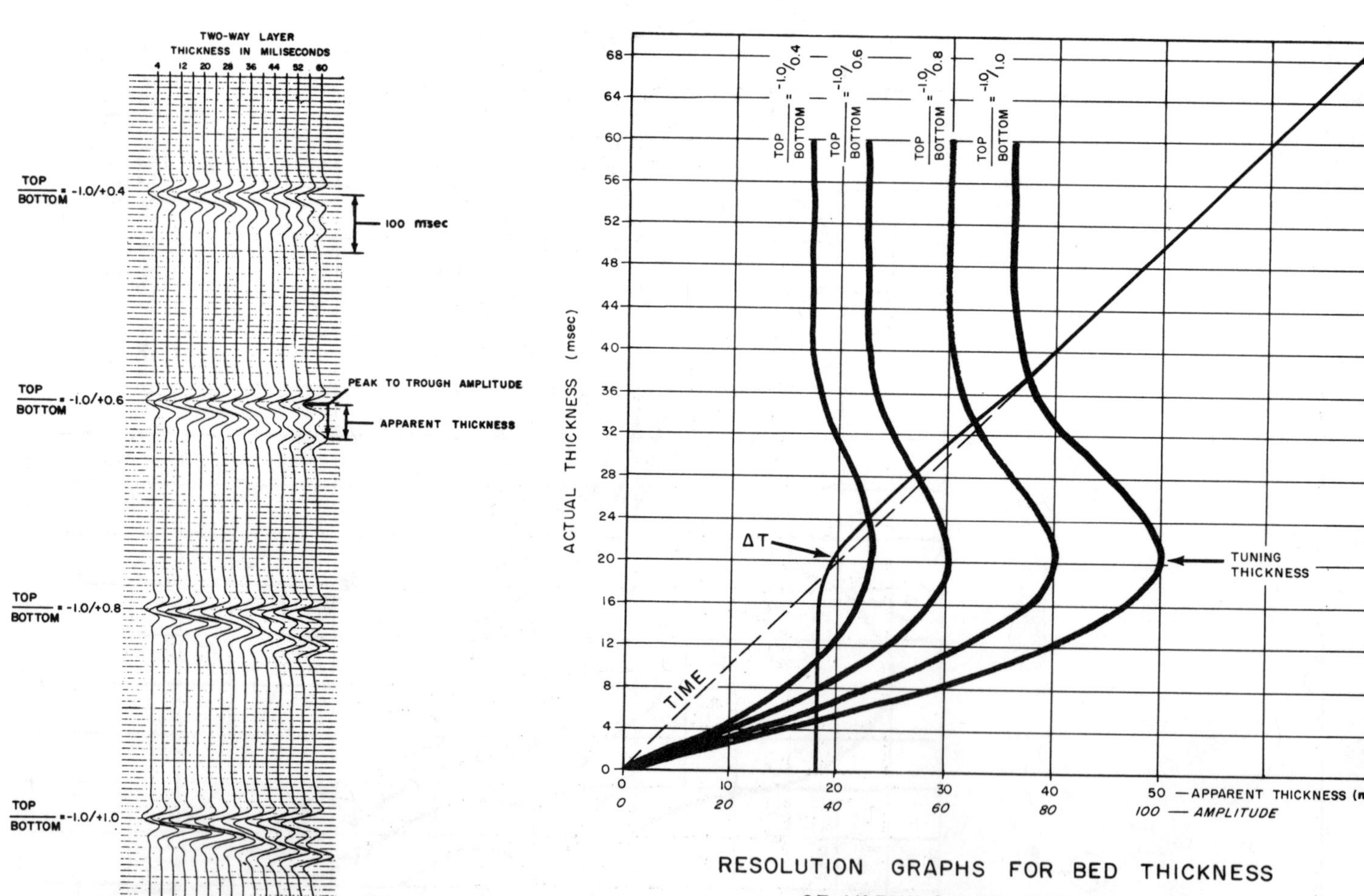

RESOLUTION TEMPLATES OF BEDS OF VARYING TOP AND BOTTOM REFLECTION STRENGTH RATIO

RESOLUTION GRAPHS FOR BED THICKNESS OF VARYING TOP AND BOTTOM REFLECTION STRENGTHS

(BAND-PASS WAVELET 8-32 HZ)

—Graph showing change in amplitude response due to change in reflection coefficients at boundary of unit.

Pursuing this theme still further, we note that porosity changes in a uniform thickness carbonate imbedded in shales again forms a circumstance where this method is applicable, but now porosity would relate to amplitude and not bed thickness. One may further consider the case where for a limestone bed in shale both bed thickness and porosity vary now making for a most challenging interpretive task.

Where we deal with intermediate acoustic impedance beds in some other lithologic scenario we must appeal to synthetic seismogram studies in order to develop an appropriate interpretive technique. Each new geologic setting may require the customized formulation of a new approach.

An example illustrating use of seismic amplitudes for a quantitative interpretation of thin bed sandstone stratigraphy is now presented in great detail so that methods and specific procedures may be noted for this case. Actual sections could not be released so models portray the various seismic profiles. The discussion which follows is taken from the contribution by Schramm, Dedman and Lindsey to the AAPG Monograph 26, Seismic Stratigraphy (1977).

"The Following case study is an actual example of the practical use of stratigraphic modeling and interpretation as it applies to exploration for and quantification of a gas-filled lens of sandstone. All of the concepts and techniques previously discussed were employed in the evaluation of this sandstone lens, lying in the offshore Louisiana portion of the Gulf of Mexico. This general location and the fact that we are dealing with Pleistocene sandstones are the only identification permitted at client request.

The prospect area includes two wells, the first of which was drilled on a structural closure (Fig. 23) and the second of which was drilled in a downdip position on the south side of a regional down-to-the-coast fault. Well No. 1 encountered a 100-ft (30.5 m) silty shale section at about -4,780 ft (1,457 m). Well No. 2 encountered a silty zone at about -5,350 ft (-1,631 m) and a 10-ft (3 m) gas sandstone at about -5,400 ft (-1,646 m). Available to the original interpreters of the prospect were standard AGC sections for the north-south lines N-1 and N-2, and for the east-west lines E-1, E-2, and E-3, plus some additional old coverage in the area of well No. 1 which is irrelevant to this problem.

In an attempt to define the position, extent, and ultimately the thickness and potential of a possible large gas-bearing sandstone lens, the data for the N and E lines were obtained, a wavelet was extracted, and the lines were reprocessed for wavelet correction. Concurrently, shooting of line SE-1 was performed for the purpose of correlating the two wells and calibrating seismic with geologic data at well No. 2. Prominent amplitude anomalies are immediately apparent on lines N-2, E-2, and E-3, and they obviously represent one coherent anomalous stratigraphic unit which is cut by a fault. The largest or most extensive segment of the anomaly occurs on the undrilled, upthrown side of the fault.

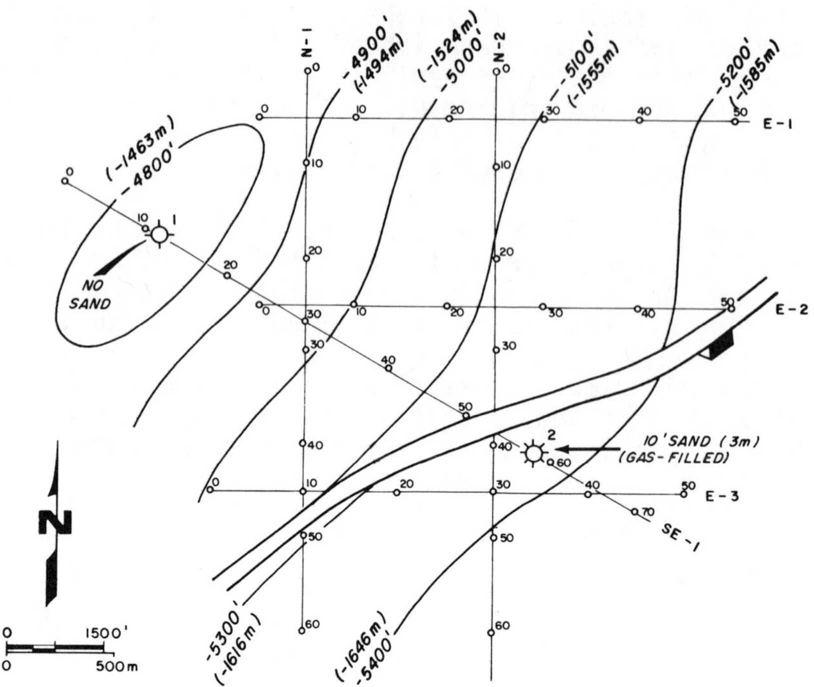

FIG. 23—Structure map, top of silty shale section.

Although the existence of an anomaly is obvious from these data, it was not obvious from the original seismic data. Relative-amplitude sections would have shown the anomaly, but would not have provided the best wavelet and, hence, information necessary to determine with assurance the existence of gas or parameters necessary for quantification.

All of the information necessary for interpretation has been condensed for each seismic line and displayed in Figures 24 through 29. Figure 31 is the contoured net-producing-zone isopach superposed on a structural base which resulted from the interpretation of these data.

Figures 25 through 29 each contain (1) a depth cross-section display of the object horizon, (2) a portion of the seismic cross section which is in true amplitude and has been wavelet-processed, (3) an expanded display of the seismic event (3) an expanded display of the seismic event which has been extracted from the complete traces for which peak-to-trough time differentials, ΔT, and relative amplitudes have been automatically measured, and (4) a display of the measured time and amplitudes from (3).

Six seismic lines (Fig. 23) were used to make this interpretation, two north-south, three east-west, and one southeast-northwest line. Interpretations have been made of each line and annotated on the amplitude ΔT plots. Two lines, N-1 and E-1, showed no abnormal amplitudes and were assumed to indicate the absence of a gas sandstone. The remaining lines did show high amplitudes and were used to measure gas-sandstone thickness. For demonstration purposes, line SE-1 (Fig. 29) will be used and the logic of interpretation explained.

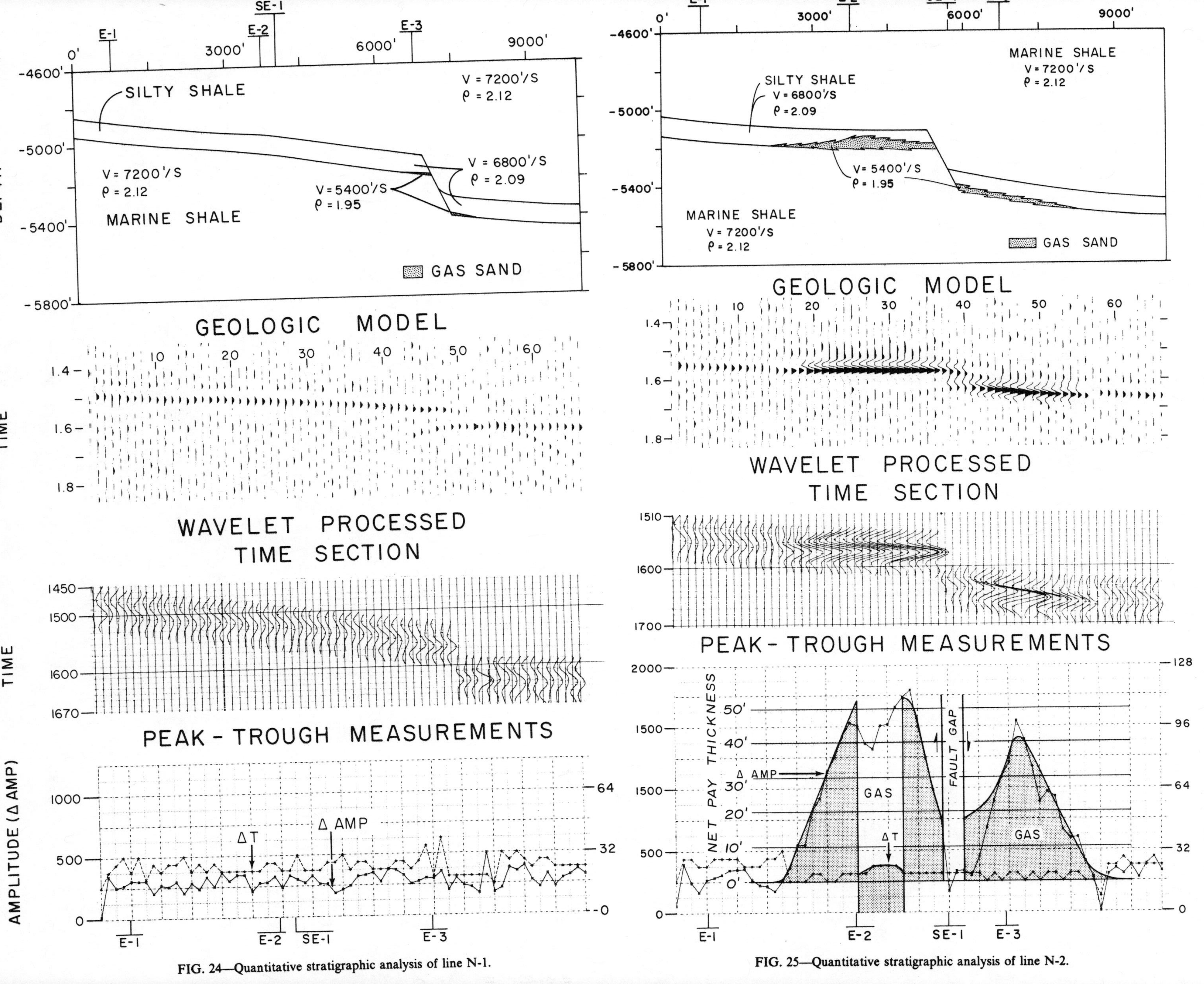

FIG. 24—Quantitative stratigraphic analysis of line N-1.

FIG. 25—Quantitative stratigraphic analysis of line N-2.

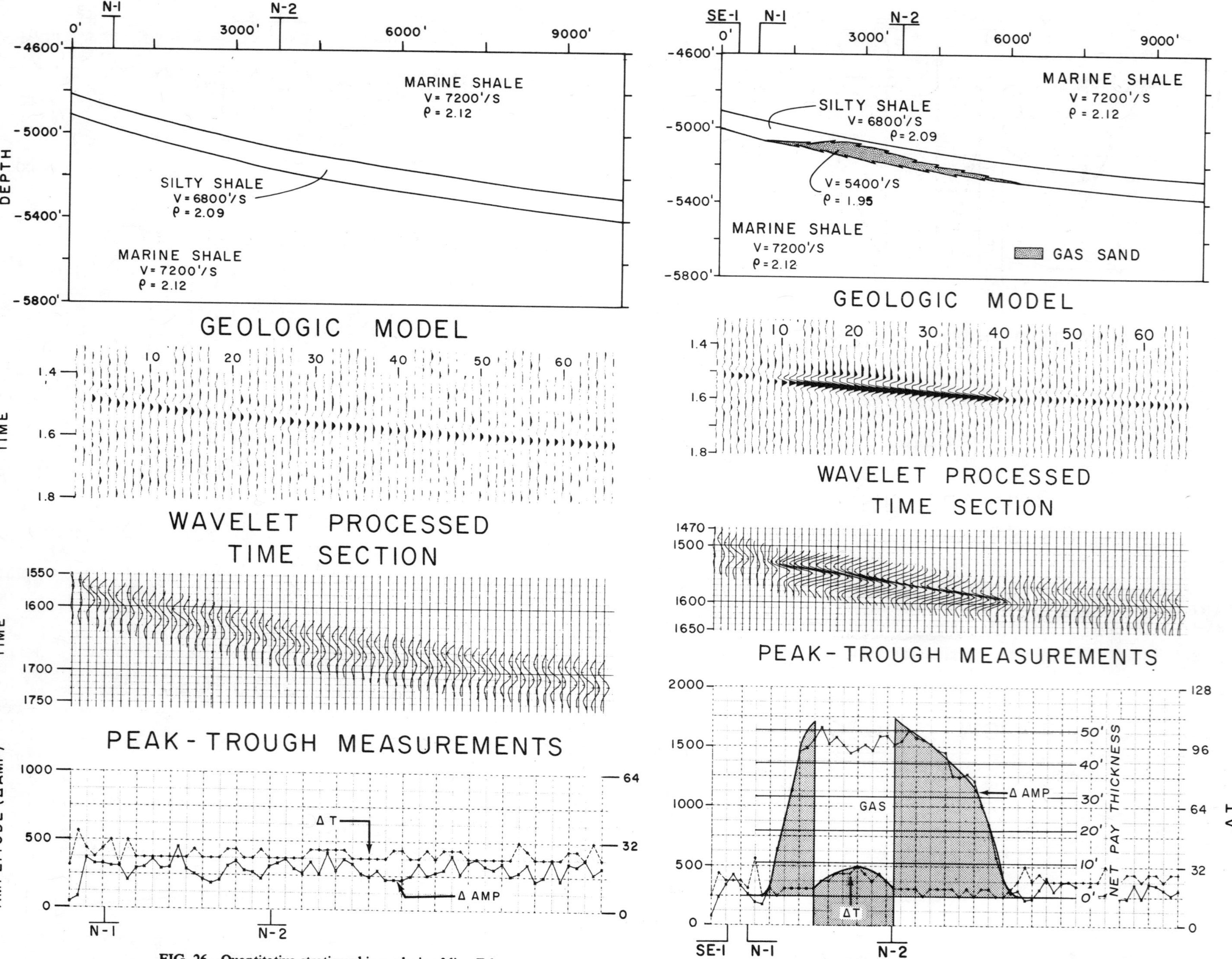

FIG. 26—Quantitative stratigraphic analysis of line E-1.

FIG. 27—Quantitative stratigraphic analysis of line E-2.

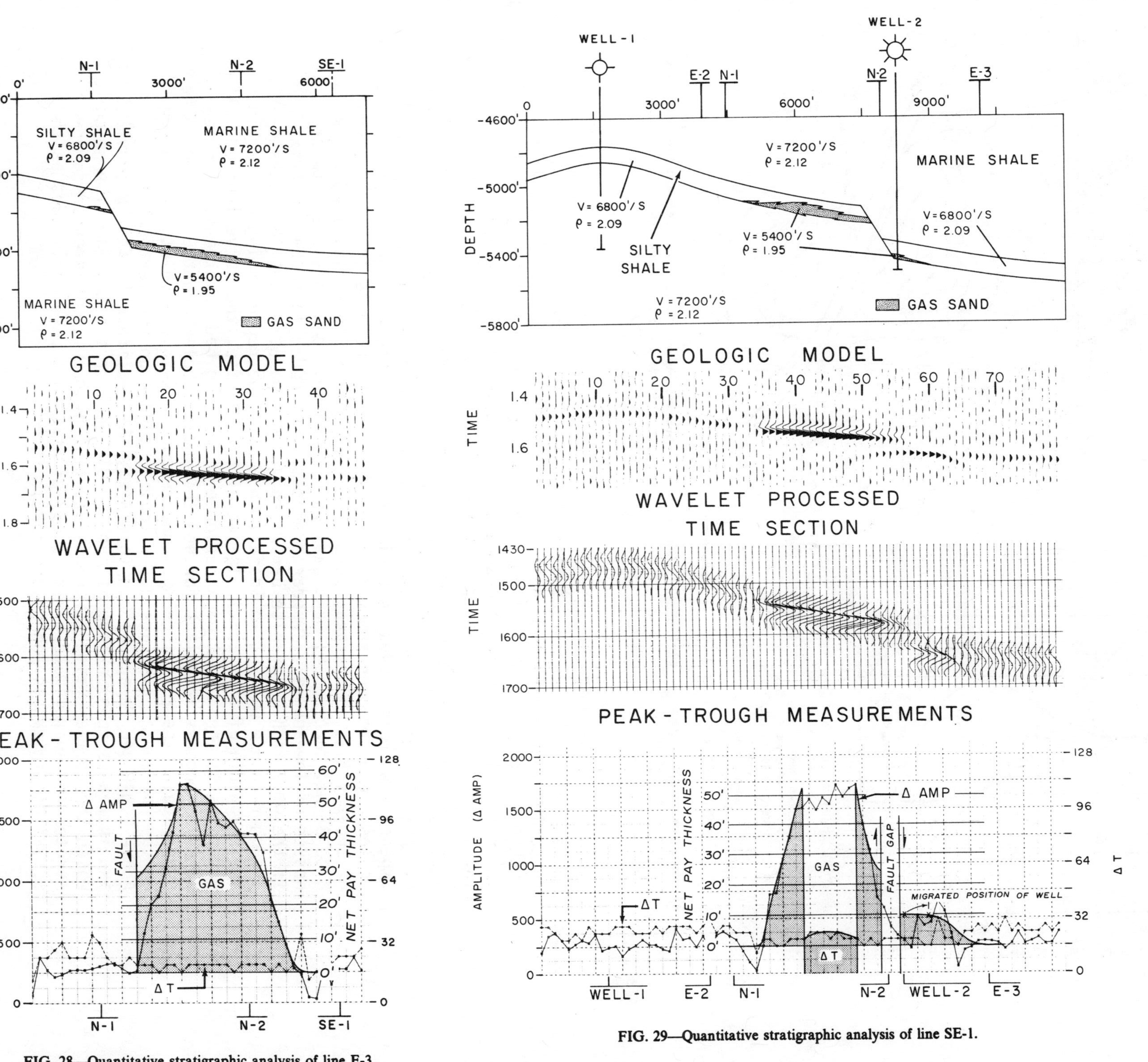

FIG. 28—Quantitative stratigraphic analysis of line E-3.

FIG. 29—Quantitative stratigraphic analysis of line SE-1.

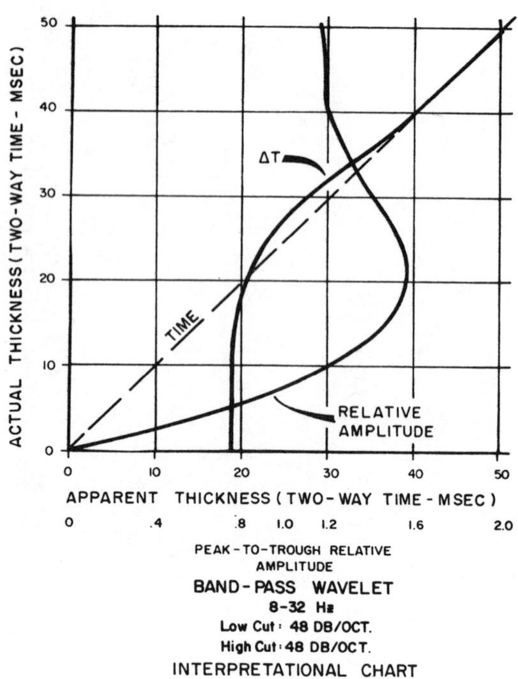

FIG. 30—Band-pass wavelet response to low-impedance thin bed.

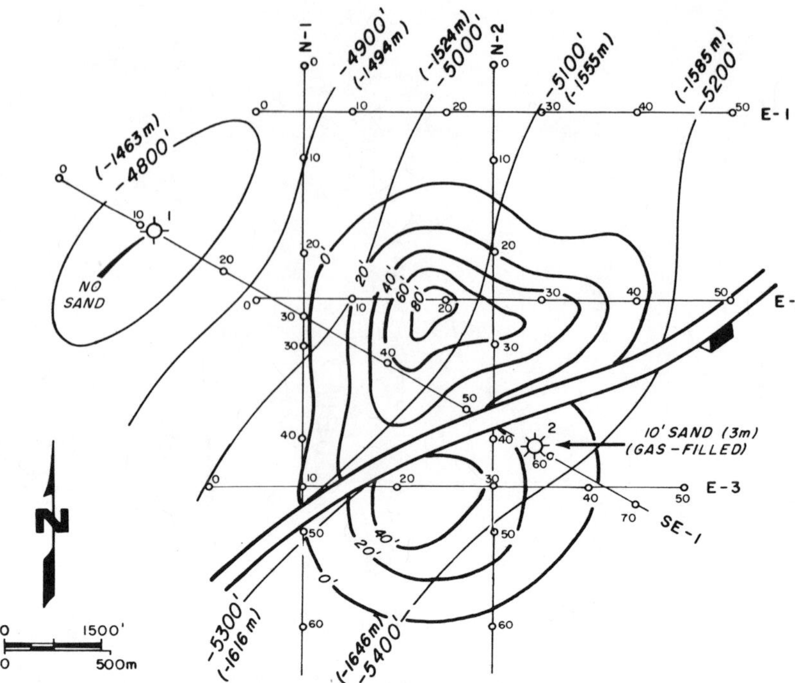

FIG. 31—Net-pay thickness map on structural base interpreted from well data and wavelet-processed seismic data.

Well 2 was drilled and showed the presence of 10ft (3 m) of net gas sandstone. The migrated position of the well was accounted for, as noted, to calibrate the amplitude data properly. Fault locations for both the upthrown and downthrown horizon positions were made and the amplitude curve and ΔT plots were hand-smoothed. A calibration plot was made for the propagating wavelet used in the wavelet processing (Fig. 30) to determine the tuning point and amplitude decay as a function of bed thickness (milliseconds). Two separate curves are presented in Figure 30; the vertical axis is actual two-way time thickness in milliseconds for low-acoustic-impedance sandstone beds in a higher acoustic-impedance shale matrix. Two scales are plotted on the horizontal axis. One scale is in milliseconds of two-way time of the peak-to-trough time separation as measured on an 8 to 32-Hz, zero-phase, band-pass wavelet convolved with varying thicknesses of low-acoustic-impedance sandstone embedded in a shale matrix. If the wavelet had infinite bandwidth, time separations of the peak-to-trough response wavelet would fall on a 45° line representing time. Since the bandwidth of this particular wavelet is finite, the curve labeled "ΔT" describes the peak-to-trough response to varying bed thickness. The measured ΔT becomes invariant for bed thicknesses below about 0.018 sec. The other scale, the relative-amplitude response of this wavelet, is plotted on the horizontal axis against actual thickness on the vertical axis. Above 0.040 sec in actual bed thickness, the amplitude is fairly constant and increases to a maximum at about 0.018 sec, or the two-way time below which the apparent time thickness becomes invariant. This curve demonstrates the approximate linear decay of peak-to-trough amplitudes as the actual bed thickness approaches zero.

"An examination of the data on SE-1 after smoothing indicates the existance of two tuning points, one at shotpoint 40 and another near shotpoint 50. The ΔT plot shows an increase between these two points which indicates that the objective net pay is resolved in time. Notice that this ΔT is less than the ΔT measurements of the silty shale bed on either side of the high-amplitude data. This is due to the fact that a reflection is being generated from the top of the gas sandstone. A phase inversion also indicates this abrupt change from a weak negative reflection at the top of the gas sandstone. Measurements of ΔT are used in the interval between shotpoints 40 and 50 as shown on Figure 29. The gas-sandstone velocity, as extracted from the sonic log of Well No. 2, of 5,400 ft (1,646 m) per second was multiplied by the tuning thickness (0.020 sec) to compute a thickness of 54 ft (16.5 m). These two values of 10 ft (3 m) and 54 ft (16.5 m) were used to construct a linear scale for calibrating the amplitude values in feet. The thickness values using ΔT's are simply computed by multiplying one-half the ΔT by the gas-sandstone velocities. The base line for the zero, net gas-sandstone thickness is chosen where the amplitude drops to that of the silty shale, which occurs at an approximate relative amplitude of 250.

The interpretation is completed by tying loops on amplitude values, where possible, and/or using the same amplitude scale in feet, as presented in Figure 29. The final interpreted map of the gas sandstone is shown in Figure 31. The thicker part of gas sandstone exists to the southwest of Well No. 2, where it is interpreted to exceed 40 ft (12.2 m) in thickness on the downthrown block, and on the north side of the fault, where it is interpreted to attain a thickness of 80 ft (24.4 m). If all other economic parameters and conditions were favorable, well locations could readily be made to test both segments of the lens, and the decision could be made as to where to set the production platform.

Amplitude measurements, as shown by this example, give the geologist and geophysicist an additional parameter for use in stratigraphic mapping in both exploratory and development projects. However, interpretation and experience are still very important ingredients in the use of these new data."

While these last discussions seem most encompassing in the interpretational technology and particularly the stratigraphic consideration employed, there is still more to the subject than delineating thin beds. We must in particular cite the pioneering work of Harms and Tackenberg (1972), Sangree and Widmier (1974), and Vail, Mitchum, Todd and Sangree (1976) who set out to achieve stratigraphic objectives by identification of sedimentary processes from seismic patterns. Significant works developed from this theme in fact form the bulk of AAPG Monograph 26, Seismic Stratigraphy (1977). Spatial and temporal resolution in concert with these approaches offers a comprehensive and most powerful technology for stratigraphic interpretation in a more complete sense. Let us now look at yet one more powerful seismic data processing procedure and the interpretation technology which follows on from it.

VIII. Conversion of Seismic Data to Log-Like Displays

A. Correlation of Seismic Data and Subsurface Measurements

When relating seismic data to stratigraphy was first discussed we noted a data comparison over a North Sea oil field in conventional and wavelet processed form (p. 20). Also, the wavelet processed data was tied to well data by means of a synthetic seismogram (p. 6). We noted an improvement in resolution in the wavelet-processed data and also the lack of multiples compared to the conventionally-processed data. This was due to the fact that the greater resolution in the data improves the velocity analysis results so that they can better discriminate between true event velocities and multiple velocities.

The wavelet used in the synthetic seismogram was the same wavelet that was in the wavelet-processed data. The synthetic seismogram was spliced into the seismic data. Excellent correlation between the synthetic seismogram and the wavelet-processed data was noted. Contrasting this to the match between the synthetic seismogram and the conventionally-processed data we concluded that using the same wavelet in both the seismic data and the synthetic seismogram is necessary in order to obtain the best possible agreement. A later discussion concerning qualitative stratigraphic correlations re-enforced this point (see p. 83).

The next Figure shows a part of a sonic log whose vertical scale has been converted to time. Note that it consists of a rapidly-varying component (high frequency) and a slowly-varying component (low frequency) generally increasing with time. The slowly-varying component, which we shall call the trend, is made up of frequency components too low to be carried by seismic data (typically below five Hertz); therefore, this information must be put into the system by some source other than seismic amplitudes. One possible source is to project the trend from a well which is close to the seismic data down the line, and if there are several wells, interpolate in between. A second method is to get this trend over a gross interval by using conventional seismically-derived velocities. The final synthetic sonic log then is the addition of the trend and the results from the inverse model, and we call these synthetic logs SYNLOGS.*

In conversion to log-like format we must then recognize that the low frequency trend and the high frequency data above the limits of the seismic sampling interval will not be recoverable. We can explicitly recognize the nature of this information loss by noting the logs from the North Sea data which has just been examined.

The following figure shows comparable portions of both a reflectivity series and an acoustic impedance log developed in two-way travel time from the North Sea velocity and density logs. Of course log editing procedures have been applied first. A filtered version of the acoustic impedance log is also shown where the frequency content is comparable to what we might expect to recover from seismic data. This then now exemplifies our correlation criterion.

*Trademark GeoQuest Int'l. Inc.

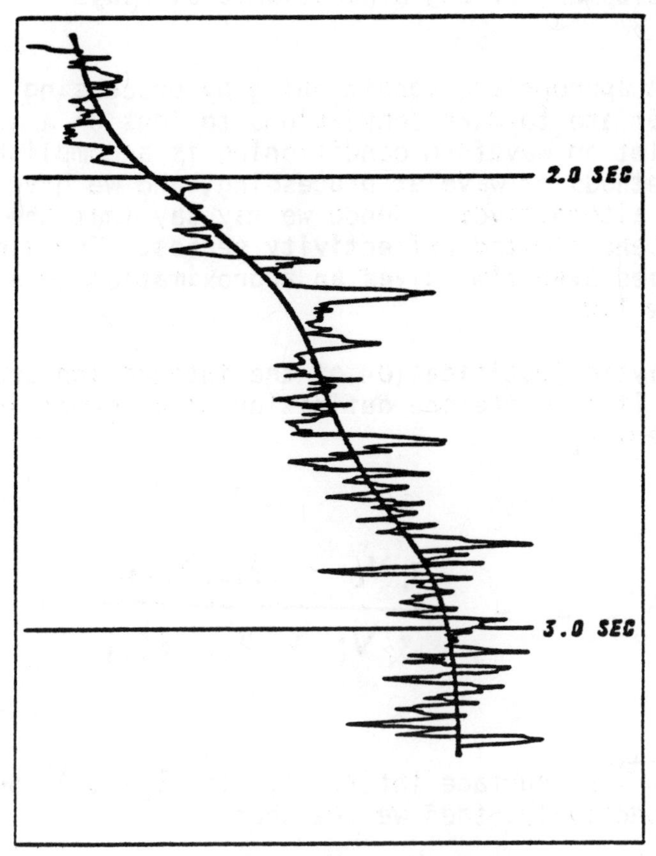

- Sonic log showing trend.

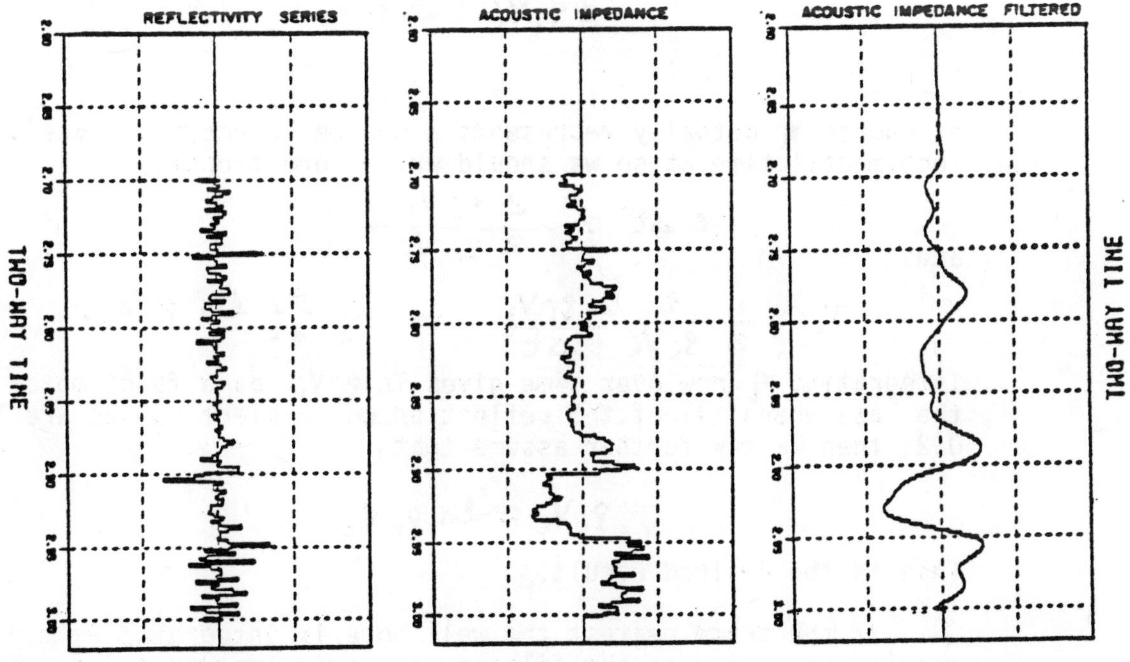

- LOG DATA - ACOUSTIC IMPEDANCE.

B. Development of Log-Like Seismic Displays

With appropriate conditioning by processing we can suggest that it is appropriate to make conversions to logs on a trace-by-trace basis. The wavelet or waveform conditioning is accomplished either by deconvolution methods or wavelet processing, and we have already given comment on these alternatives. Hence we may say that the seismic data constitutes a band-limited reflectivity series. The simple integration of this series over time gives an approximation to a filtered acoustic impedance log.

Analytic justification of the integration process is readily demonstrated. If we write the definition of a normal incidence reflection coefficient r_i

$$r_i = \frac{\rho_i V_i - \rho_{i-1} V_{i-1}}{\rho_i V_i + \rho_{i-1} V_{i-1}}$$

for the ith subsurface interface, with ρ_i and V_i being density and velocity respectively, then we see that

$$r_i = \frac{\Delta \rho_i V_i}{2\rho_i V_i + \Delta \rho_i V_i} \approx \frac{\Delta \rho_i V_i}{2\rho_i V_i} .$$

of course r_i actually represents a change in acoustic impedance over an incremental time Δt so we should write more properly

$$r_i \Delta t \approx \frac{\Delta \rho_i V_i}{2 \rho_i V_i}$$

and

$$r_i \approx \frac{1}{2} \frac{1}{\rho_i V_i} \frac{\Delta \rho_i V_i}{\Delta t} = \frac{1}{2} \frac{d}{dt} \ln \rho_i V_i .$$

Integrating r_i now over time gives $\ln \rho_i V_i$ as a function of time. In the case where all of the reflection coefficient values are less than 0.2, then we may further assume that

$$\rho_i V_i \approx \ln \rho_i V_i .$$

This is the desired result.

If the trace nearest the well bore is integrated we may compare the result produced with the filtered acoustic impedance log. This comparison is shown next. The agreement is excellent. We can clearly note the Jurassic sand which is the pay zone and the overlying low acoustic impedance black organic hot shale.

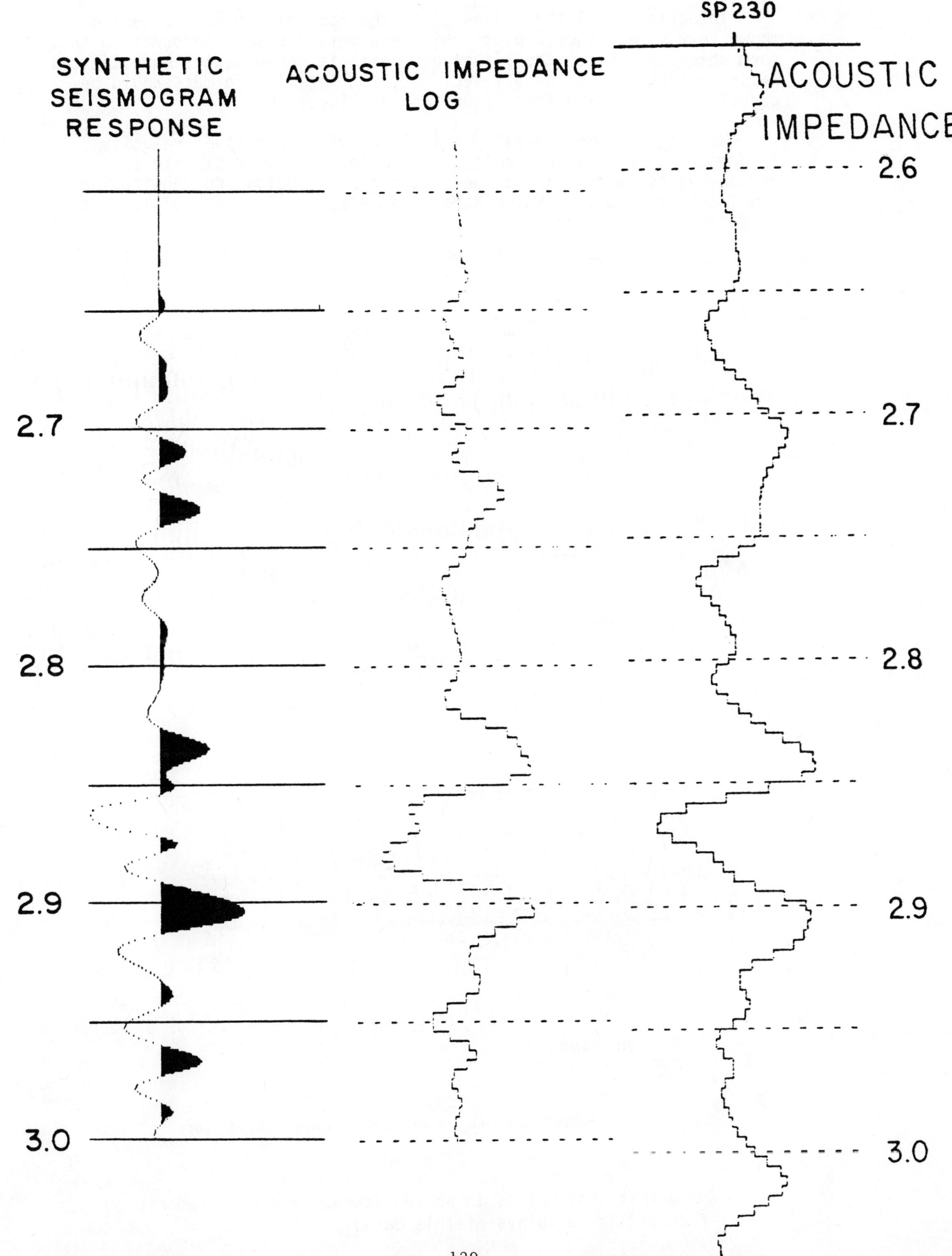
SYNTHETIC
SEISMOGRAM
RESPONSE
ACOUSTIC IMPEDANCE
LOG
SP 230
ACOUSTIC
IMPEDANCE
2.6
2.7
2.8
2.9
3.0
2.7
2.8
2.9
3.0

Integrating all the traces gives the acoustic impedance section shown next. The "square plot" which presents the same information in a form which emphasizes the lithologic units is shown next. Note the oil-water contact just below 3.1 sec. Data in log-like format should be correlated as if each trace represented a log.

We should comment finally that if we wish to recover velocity information rather than acoustic impedance, then we must make some reasonable assumption about the relation of velocity and density. One usual approach is to involve Gardner's relation (Gardner 1974), which leads to

$$\rho \propto V^{1/4}$$

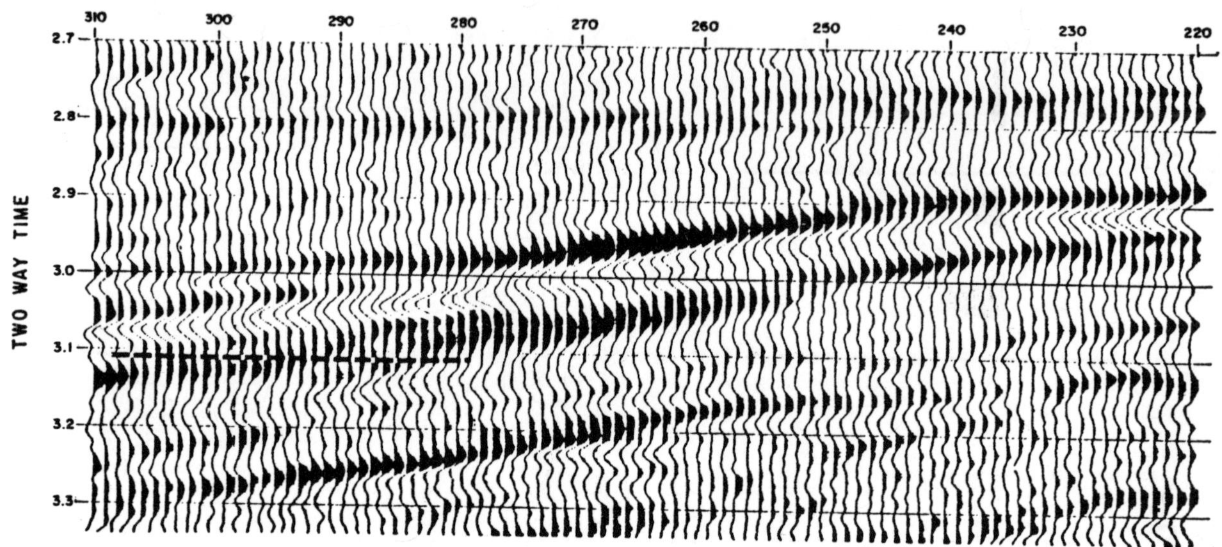

- North Sea line 1 - acoustic impedance logs.

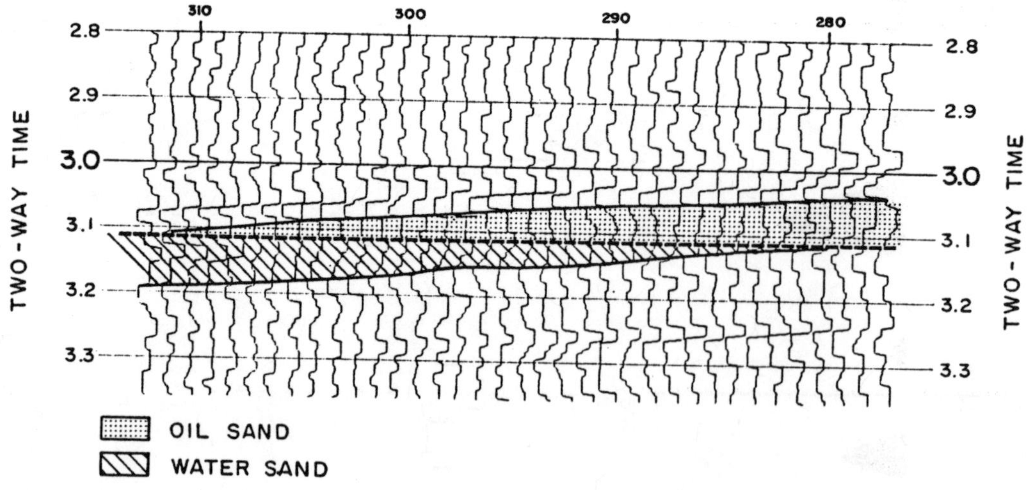

- North Sea line 1 - acoustic impedance square-plots.

(Note that shot points in above Figures have been renumbered from earlier displays of this data)

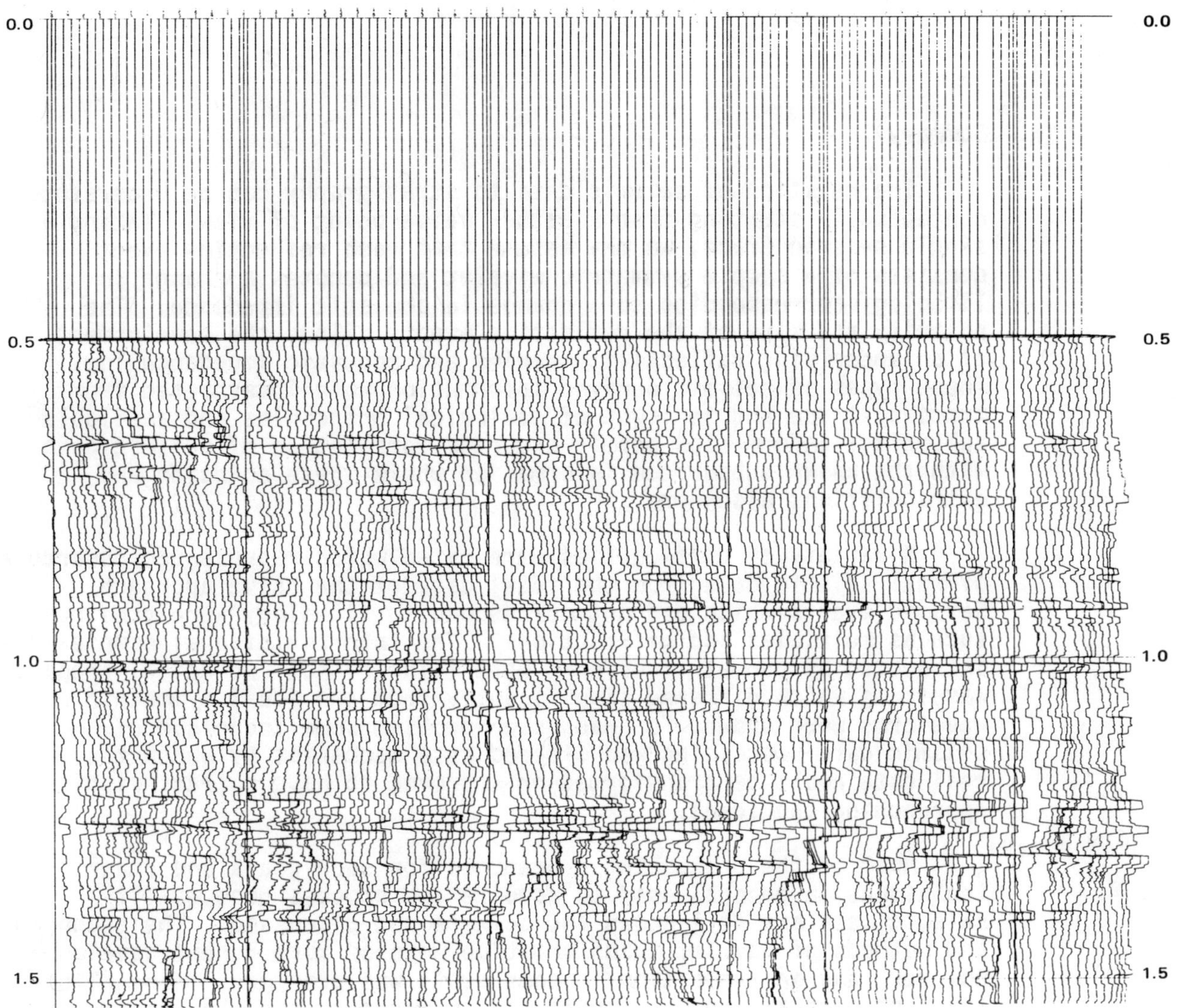

Internal Velocity Estimates (Central Oklahoma)

Courtesy of Seismograph Service Corp.

The Figure above shows interval velocity estimates for the Central Oklahoma line which was wavelet processed by Seismograph Service Corporation and shown earlier (p. 49). We should mention in particular the work of Lavergne and Willm (1977) and Lindseth (1976,1977) who were pioneers in the development of this technology. Lindseth further recognized the value of color displays in presenting subsurface parameter estimates and this is rapidly becoming an industry standard.

IX. Summary

The relation of seismic data to stratigraphy can be understood conceptually starting from a rather naive point of view. Taking a single seismic trace, we may assume that each reflection event consists of a simple symmetric wavelet and denotes a change in lithology corresponding to a change in acoustic impedance. The polarity and size of the reflection events allows the development of an underlying reflectivity series and acoustic impedance log. These types of seismic derived information when interpreted from trace-to-trace and in light of geological principles and available subsurface information permit accurate correlation of seismic data with the known subsurface information where these coincide and extrapolations elsewhere.

A variety of difficulties were cited which represented departures of real seismic data from the ideal data of the naive approach. Seismic events which did not correspond to lithology were considered such as multiples and diffractions. The interpretational complexities introduced by geometric effects were illustrated, and actual propagating waveforms which differed significantly from the ideal symmetric waveforms were shown. For each such problem some technique of seismic data processing was offered as a mechanism by which the real data could be transformed or improved to have characteristics approaching the ideal.

Most of the processing procedures for such goals and other techniques in general used to acquire and process good quality seismic data are well known and so were not discussed. The technology and methods of wavelet transformation were of more recent development however, and an exposition in some depth was provided. Many illustrations emphasizing the importance of waveform and the utility of the symmetric waveform were provided. A knowledge of the propagating waveform was recognized as a fundamental requirement for wavelet processing.

A specific vehicle by which correlation between seismic data and stratigraphy was to be effected is seismic modeling. Seismic modeling was understood to be the successor to the synthetic seismogram calculation which is in fact the most elementary form of earth model. Illustrations of modeling using ray trace procedures and wave theory calculations in two and three dimensions brought forth a variety of considerations of profound interpertational significance. Quite apart from re-emphasizing the role of seismic waveform we learned that lateral continuity of lithology over the horizontal extent of a Fresnel zone is a hard and fast requirement for the correlations we seek to observe.

Another lesson from the exemplary model studies warned the interpreter to constantly account for the three dimensional nature of the subsurface. Further, the great instructive power of modeling particular exploration objectives was clearly demonstrated and the practical successes deriving from such insights were documented.

Next, qualitative stratigraphic correlations were considered based on a variety of schematic models and real examples. Amongst the circumstances treated were lateral lithologic transitions, delineation of differing depositional environments for the same materials, lithologic transitions in the presence of structure and certain of the reservoir engineering implications from the correlations we are able to make. The studies made it clear that a principal effect of transitioning was a reduction in seismic amplitude levels and that quite refined correlation between seismic data and geology were possible. In fact, the refinement or resolution depended very greatly upon the geological information and insights which could be brought to bear and these were applicable even in the presence of structure. The final illustration of the discussion which centered about a lithologic transition on the flank of an anticlinal structure pointed out the folly which would result if stratigraphic considerations had been made subordinate to structural ones.

Finally, the stratigraphic resolution potential of seismic data was expressed in quantitative terms. A computer based procedure was developed which used both time measurements and amplitudes in the amplitude processed, wavelet processed environment to achieve high resolution lithologic delineation. The method was explained along with necessary calibration procedures and illustrated using schematic and real data cases. At the same time, processing of seismic data to recover filtered log-like information was noted giving an alternative mechanism for making correlations.

The future for stratigraphic interpretations of seismic data and finding hydrocarbons in stratigraphic environments seems particularly bright at this time, particularly in the light of available tools and methods. Amplitude and wavelet processing, migration, geoseismic modeling, recovery of log-like parameters, and quantitative methods for delineating thin bed lithology are some of the tools which have carried us to this plateau of progress. On the horizon are tools which will process logs and seismic data in concert to recover probabilistic lithologies directly. We must watch with keen interest as newer tools and procedures from both geology and geophysics further improve our capabilities.

X. Acknowledgements

These notes represent a distillation of an approach to stratigraphic interpretation developed over a number of years by several individuals at GeoQuest International, Ltd. It is important that these contributions be recognized as most of the discussions document their work. In particular we must cite J. P. Lindsey, E. V. Dedman, Dr. M. W. Schramm and E. Poggiagliolmi. A second category of individuals assisted in preparation of studies and these efforts as well should be recognized.

XI. Bibliography

AAPG, Seismic Stratigraphy - Applications to Hydrocarbon Exploration, Memoir 26, Ed. Payton (C. E.), pp. 516, Tulsa, 1977

Berkhout (A. J.), "On the Minimum-Lenth of One-Sided Signals", Geophysics, Vol. 38, pp. 657-672, 1973
"Related Properties of Minimum-Phase and Zero-Phase Time Functions", Geophys. Prosp., V. 22, pp. 683-703, 1974

Biggs, (M), Blakely, (R), and Rudman, (A), "Seismic velocities and a synthetic seismogram computed from a continuous velocity log of a test well to the basement complex in Lawrence County, Indiana," Indiana Department Consv., Geol. Survey, Report of Progress No. 21, 15 p., 1966

Dedman (E. V.), Lindsey (J. P.) and Schramm (M. W.), "Stratigraphic Modeling: A Step Beyond Bright Spot", World Oil, V. 180, No. 6, pp. 61-65, 1975

Frink (A. P.), Wittick (T. R.) and Dedman (E. V.), "Interpretive Methods of Stratigraphic Modeling", Preprint, OTC, Houston, May 2-5, 1977

Gardner (G. H. F.), Gardner (L. W.) and Gregory (A. R.), "Formation Velocity and Density - The Diagnostic Basics for Stratigraphic Traps", Geophysics, V. 39, No. 66, pp. 770-780, 1974

Harms (J. C.) and Tackenberg (P.), "Seismic Signatures of Sedimentation Models", Geophysics, V. 37, p. 45-48, 1970

Harms, (J. C.), and Tackenberg, (P), "Seismic signatures of sedimentation models", Geophysics, V. 37, No. 1, p. 45-58, 1972

Hilterman, (F. J.), "Three-dimensional seismic modeling" Geophysics, V. 35, No. 6, p. 1020-1037, 1970

Geophysical Society of Houston, A symposium: lithology and direct detection of hydrocarbons using geophysical methods, Houston, October 8-9, 1973

Jolly, (R. N.), and Mifsud, (J. F.), "Experimental studies of source-generated seismic noise," Geophysics, V. 36, No. 6, p. 1138-1149, 1971

Lavergne (M.) and Willm (C.) "Inversion of Seismograms and Pseudo Velocity Logs" Geophys. Prosp., V. 25, No. 2, 1977

(Bibliography (cont'd.)

Lindseth (R. O.), "Seislogs", Lecture Notes, AAPG-SEG School: Stratigraphic Interpretation of Seismic Data, Houston, Texas, Sept. 20-25, 1976

-----"An Improved Method for Offshore Exploration", Preprint OTC, Houston, May 2-5, 1977

Lindsey, (J. P.) and Craft (C. I.), "How Hydrocarbon Reserves are Estimated from Seismic Data", World Oil, August 1, 1973

----- Schramm (M. W.) and Nemeth (L. K.), "New Seismic Technology Can Guide Field Development, World Oil, V. 183, No. 7, pp. 59-63, 1976

McClintock (P. L.), "Seismic Data Processing Techniques for Northern Michigan Reefs", AAPG Stud. Geol., No. 5, 1978

Meissner (R.) and Meixner (E.), "Deformation of Seismic Wavelets by Thin Layers and Layered Boundaries", Geophys. Prosp., V. 17, pp. 1-27, 1975

Morrison (O. J.), "A Discussion of the Features of the AIMS Modeling System", Brochure, Seismograph Service Corp., 1975

Nath (A. K.), "Reflection Amplitude, Modeling Can Help Locate Michigan Reefs", Oil & Gas Journal, V. 73, No. 11, pp. 180-182, 1975

----and Patch (J. R.), "Full Utilization of Seismic Resolution - The Key to Deterministic Stratigraphic Interpretation from Marine Data", Preprint, OTC, Houston, May 2-5, 1977

Neidell (N. S.), "What are the limits in specifying seismic models?" Oil and Gas Journal, February 17, 1975

-------and Taner (M. T.), "Semblance and other Coherency Measures for Multichannel Data", Geophysics, V. 36, No. 3, pp. 482-497, 1971

Peterson (R. A.), Fillipone (W. R.), and Coker (F. B.), "The synthesis of seismograms from well log data" Geophysics, V. 20, No. 3, p. 516-538, 1955.

Ricker (N.), "Wavelet contraction, wavelet expansion and the control of seismic resolution" Geophysics, V. 18, No. 4, p. 769-792, 1953

Robinson (E. A.), "Predictive Decomposition of Seismic Traces", Geophysics, V. 22, pp. 767-778, 1957

-------and Treitel (S.), "Principles of Digital Wiener filtering", Geophysical Prospecting, V. 15, No. 3, p. 311-333, 1967

Sangree (J. B.), and Widmier (J. M.), "Interpretation of depositional facies from seismic data", Continuing Education Symposium Geophysical Society of Houston, December 4-5, 1974

(Bibliography (cont'd)

Schoenberger (M.), "Resolution comparison of minimum-phase and zero-phase signals", Geophysics, V. 39, No. 3, p. 600-604, 1974

Shah (P. M.), "Ray tracing in three-dimensions", Geophysics, V. 38, No. 3, p. 600-604, 1974

Sheriff (R. E.), "Inferring stratigraphy from seismic data", AAPG Bulletin, V. 60, No. 4, p. 528-542, 1976

Taner (M. T.), and Koehler (F.), "Velocity spectra - digital computer derivation and applications of velocity functions", Geophysics, V. 34, No. 6, p. 859-881, 1969

-------, Cook (E. E.), and Neidell (N. S.), "Limitations of the reflection seismic method; lessons from computer simulations", Geophysics, V. 35, No. 4, p. 551-573, 1970

Vail, (P. R.), Mitchum (R. M.), Todd (R. G.) and Sangree (J. B.), Interpretation of seismic sequences from reflection patterns," Preprint, 29th Annual Midwestern SEG, Dallas, March 7-9, 1976

Widess (M. B.), How thin is a thin bed - Geophysics, V. 38, No. 6, p. 1176-1180, 1973

Wuenschel (P. C.), "Seismogram synthesis including multiples and transmission coefficients", Geophysics, V. 25, No. 1, pp. 106-129, 1960

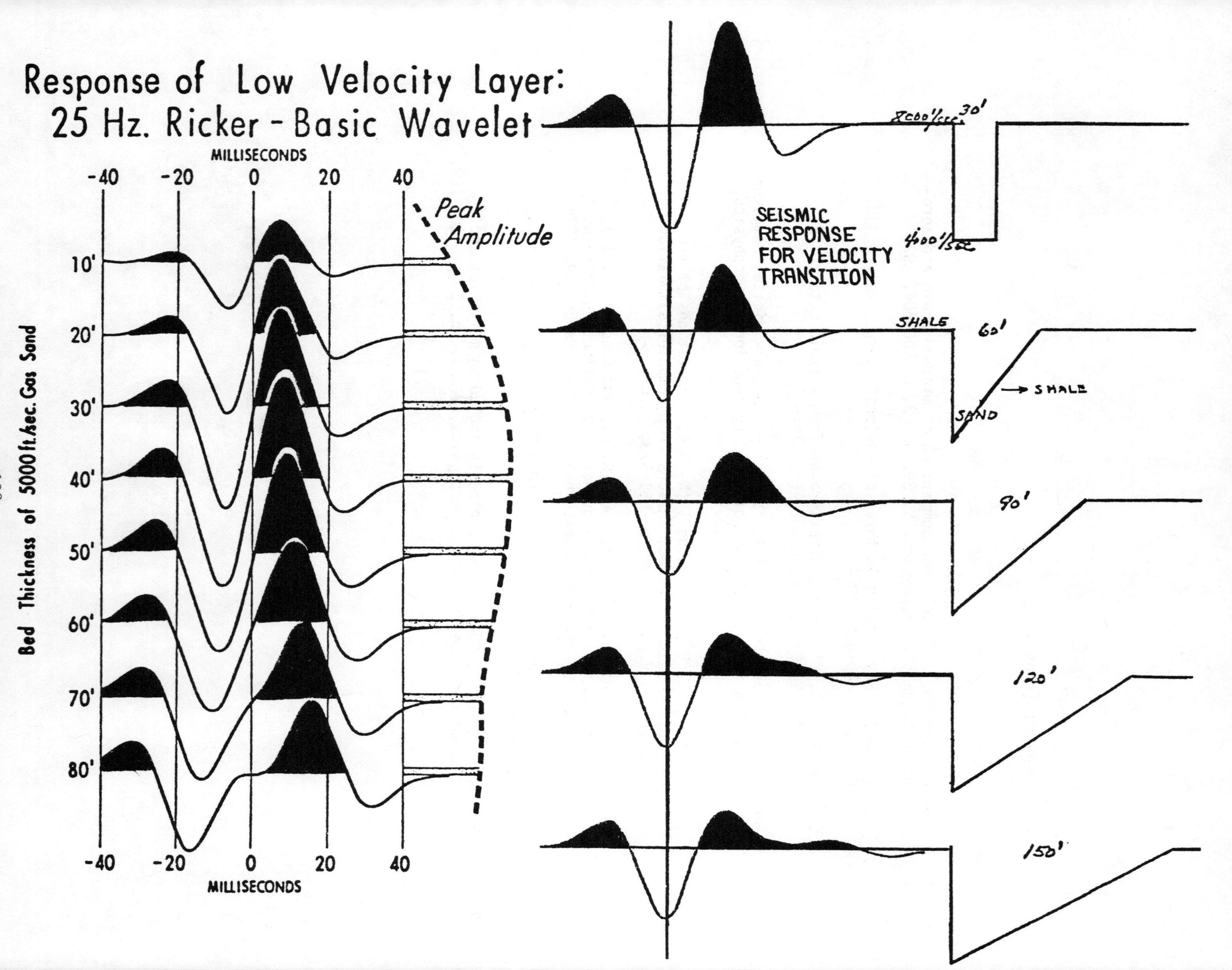

Response of Low Velocity Layer:
25 Hz. Ricker - Basic Wavelet
MILLISECONDS
-40
-20
0
20
40
Peak
Amplitude
Bed Thickness of 5000ft./sec. Gas Sand
10'
20'
30'
40'
50'
60'
70'
80'
MILLISECONDS
SEISMIC
RESPONSE
FOR VELOCITY
TRANSITION
30'
SHALE
60'
→ SHALE
SAND
90'
120'
150'

Stratigraphic Correlation
Workshop

On the facing page waveforms are shown which denote thick and thin stratigraphic units which have either sharp boundaries or else one transitional boundary. Study these and note the relations between the stratigraphic unit and the waveshape and size. Note subtle clues - is the peak bigger than the trough? - has the waveform lengthened?

The task on the two pages which follow is to match the indicated waveforms up with the lithologic models. In addressing this task, eliminate the easy ones first . . .

STRATIGRAPHIC CORRELATION
QUIZ

A.

BASIC WAVELET
(RICKER 25 HZ.)
CORRESPONDS TO
?__________(SEE
STRATIGRAPHIC MODELS

B. ____________

C. ____________

D. ____________

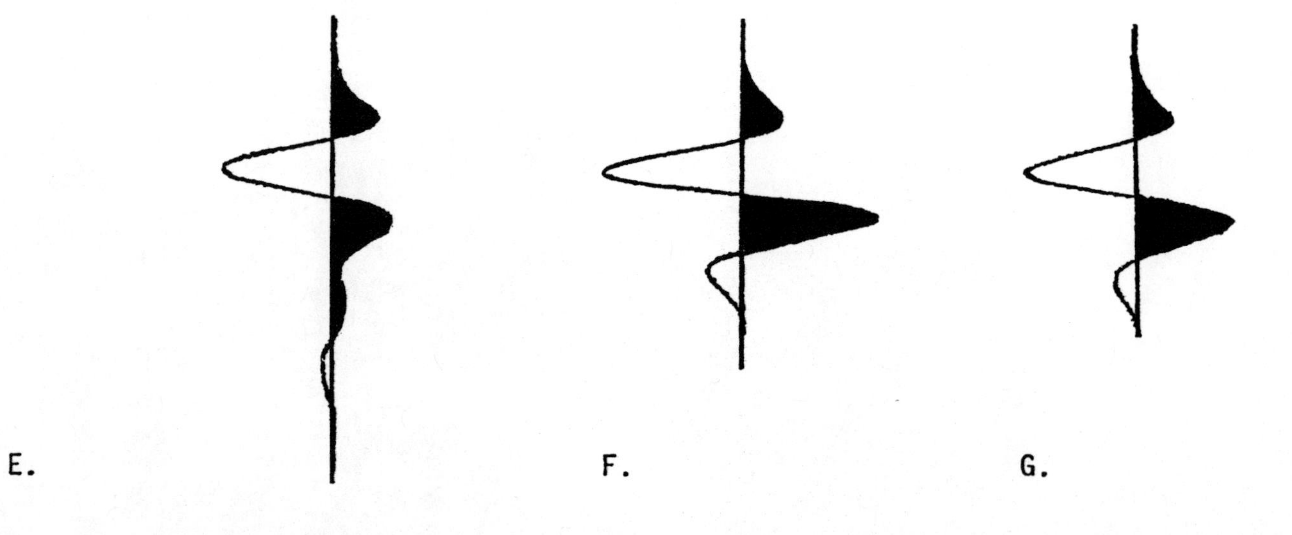

____________ ____________ ____________

SOME HINTS AND CLUES -

NOTE:
- INDICATED POLARITY OF EVENT
- SEPARATION OF PEAKS AND TROUGHS
- AMPLITUDE RELATIONS
- OVERALL LENGTH OF EVENT COMPLEX

STRATIGRAPHIC

CORRELATION QUIZ (CONTINUED)

STRATIGRAPHIC MODELS

SINGLE REFLECTING INTERFACE

		MODEL A
HIGH VELOCITY	V_H	
LOW VELOCITY	V_L	

SINGLE LOW VELOCITY LAYER

HIGH VELOCITY V_H

LOW VELOCITY V_L τ

HIGH VELOCITY V_H

τ	V_L	MODEL
10'	5000'/SEC	B
30	5000	C
30	4000	D
60	5000	E

TRANSITION ZONES

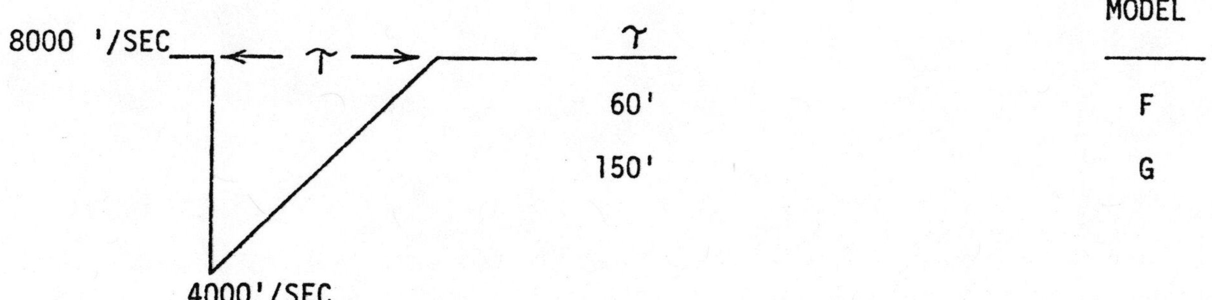

τ	MODEL
60'	F
150'	G